INSTANT

BIOLOGY

INSTANT

BIOLOGY

FROM SINGLE CELLS TO HUMAN BEINGS, AND BEYOND

BY BOYCE RENSBERGER

A Byron Preiss Book

FAWCETT COLUMBINE • NEW YORK

Copyright © 1996 by Byron Preiss Visual Publications, Inc.

All rights reserved under International and Pan-American Copyright Conventions. Published in the United States by Ballantine Books, a division of Random House, Inc., New York, and simultaneously in Canada by Random House of Canada Limited, Toronto.

Cartoon Credits: © 1995 Aaron Bacall—101. © 1995 Roz Chast—29. © 1995 Robert Mankoff—60, 181.

The Cartoon Bank, Inc., located in Yonkers, NY, is a computerized archive featuring the work of over fifty of the country's top cartoonists.

Illustration Credits: © 1995 Archive Holdings—23, 65, 67, 158, 159, 160, 162, 163, 164, 166, 169, 172, 179, 201. © 1995 Dartmouth Publishing, Inc.—20, 46, 47, 51, 54, 70, 75, 76, 93, 95, 108, 109, 110, 121. Purves, Orians, Heller, *Life: The Science of Biology, Third Edition.* © 1995 Sinauer Associates, Inc., Sunderland, MA—8, 17, 19, 40, 41, 50, 78, 79, 81, 82, 84, 88, 90, 103, 115, 116, 118, 122, 124, 126, 131, 142, 146, 185, 188, 204, 205. American Society for Cell Biology—61. Boyce Rensberger—63, 72, 73, 87, 97. David Ward, Yale Univ. School of Medicine—38.

Library of Congress Catalog Card Number: 95-90704

ISBN: 0-449-90701-5

Cover design by Wendi Carlock

Manufactured in the United States of America

First Edition: February 1996

10 9 8 7 6 5 4

20777 3607

CONTENTS

INTRODUCTION

It's Friday night and you're in the mood for a real book. Your mind is crying out for something that will satisfy its curiosity about what is probably the most astonishing phenomenon on Earth. You've read the books of essays by famous biologists, and they're fine, but they didn't really explain the facts of life to you . . . the facts of all life. You want to understand how life works, what it is that makes a hunk of matter live and breathe, move around, and, in your case, think about itself. You want to understand the mystery—ah, sweet mystery—of life itself, of life in all its diverse forms on this planet. And you want to know how it all came to be.

So you head down to the local bookstore and scan their meager biology section. What should you buy? To make sure you cover all the bases, you pick out an arm-load of titles that seem relevant, based on what you remember from high school biology. Two months later, the books remain where you left them, unread. All those technical terms are just too daunting. The dust on the stack is pretty impressive. Time passes. A lot of time. You know it's all over when you buy a book that claims plants have a "secret life" and can read your mind. You feel cheap in the morning.

Now that you've found this book, though, you can plunge right into the real stuff about the science of living things, and it'll be a pleasure. Trust us. You'll be learning about, among other things, yourself—your own body and how you fit into the grand scheme of living things. Life is fascinating and the way we cover it in this book, the fascination comes through. That's because we took out the jargon and unnecessary technical terms—many of which even biologists don't use, but biology teachers keep putting on exams. This book is written not so we can test your vocabulary but so you can understand biology. And when people really understand a subject, they love it.

WE CAN TELL: INSTANT BIOLOGY IS FOR YOU

In this survey of the main areas of biological science, we'll cover all the major discoveries and concepts in the life sciences that a distinguished panel of biologists has determined every educated person ought to know, and then some. The panel was put together by the American Association for the Advancement of Science as part of a long-term effort to reform science education in the nation's schools. The AAAS is the country's largest scientific society. Its panel wasn't trying to create junior scientists, just people who had a grasp of the key facts and ideas that shape modern science. Panel members felt that if Americans knew these concepts, they would be equipped to function as informed citizens in an increasingly science- and technology-based world, a world in which human activity is having profound effects on the biosphere. The AAAS panel published a list of key concepts for all the sciences. We adopted the biology section in determining what should be in this book.

CHAPTER 1: THE DIVERSITY OF LIFE

We start with something that's so obvious, we seldom stop to think about it. But when we do think about it, it reveals some extraordinary things about life on Earth. Namely, this place is crawling with critters of every description. There are millions of different kinds of species, so many that most of them haven't even been discovered yet. We'll explain how we know that. It doesn't make the headlines, but several new species are discovered every day. And they're not all teeny ones. In this century alone, eleven new species of whale have been recognized.

We'll see how scientists have tried to make sense out of all the biodiversity by classifying species according to similarities and differences. Sometimes the results are surprising. For example, if you learned in school that there were only two kingdoms, plants and animals, then you have a treat in store. There are now five: three more realms of living things that are neither vegetable nor animal (and they're not mineral, either).

For all the diversity, however, modern biology has discovered a fundamental similarity, one that carries a profound implication. Every living thing on Earth uses the same genetic code, the same way of arranging special molecules to store information in genes. In Chapter 2 you'll learn how that code works, but for now there is a larger point worth pondering. It is an arbitrary code. In other words, any number of other codes would work just as well. That being so, how come there is only one genetic code? Hold that thought, and its full implication will emerge in the last chapter.

CHAPTER 2: HEREDITY

From the wide world of Chapter 1, we take you directly to the smallest realm in biology: genes in this chapter and cells in the next.

Everybody knows we inherit traits from our parents. Genes are the reason, and in this chapter you'll learn exactly how they work. If you're into computers, one analogy may help. Genes are the software of life. They contain the programs that guide the processes inside cells, which are the hardware. Eye color is just one of between 50,000 and 100,000 things that genes guide in the human body.

How do genes do these things? Each one tells a cell how to make one kind of protein molecule. Proteins are the workhorse molecules of life. Some are building blocks of cells and organs; many more act as workers of various kinds in cells (enzymes, they're called).

We'll recap the story of how Gregor Mendel, a nineteenth-century Austrian monk, discovered a key fact about genes when breeding peas in the monastery garden. He deduced that inherited traits aren't a blend of features from the two parents but are passed on as discrete units, and that a unit from one parent can override (be dominant over) a unit from the other parent. We now call the units genes.

You'll learn, in more detail than you ever thought you could understand, how genes are organized into chromosomes, how a gene communicates its messages to the cell, and what the cell does in response to the gene.

And you'll come to grasp something that most people never understand. Even though genes are extremely powerful and have a lot of say—so, they are not autonomous. Genes don't do anything unless told to do so by other things, such as messenger molecules coming from other parts of the cell or from the environment outside the cell.

CHAPTER 3: CELLS

In this part we come to the heart of life itself—the fundamental unit of all forms of life, the cell. A cell isn't just a blob of protein. It is an incredibly complex machine made of thousands of different kinds of parts, all interacting in a coordinated way to take in nutrients, grow, move around, sense its environment, and even reproduce. In other words, a cell is alive. Some cells are independent organisms (you will have met them in Chapter 1), and others are members of giant colonies called, well, plants and animals. Whether a one-celled organism or a member of a multicelled colony reading a book, cells all work the same basic way.

We'll look at the history of "cell theory" and review some of the most famous experiments in ancient biology, ones that proved that all life on Earth today is descended from pre-existing forms of life. Sounds trivial but back in the centuries we think of as being so imbued with religion, Europeans were quite ready to believe that life was not so miraculous and that it was constantly popping into existence all around them.

You'll also gain a working knowledge of all the major components of the cell and, finally, we'll look at what is probably the most amazing thing cells do (other than read books)—divide or, if you prefer, multiply. Either way, you'll get an up-close look at how one unit of life transforms itself into two units of life.

CHAPTER 4: THE HUMAN BODY

Here, we get very personal. We look at your body. Well, not that personal. This is only a book. We do start with something rather intimate, though: the creation of a human life when a sperm and an egg meet in the prospective mother's fallopian tube. You'll learn, for example, that there is no one moment that can be called conception. Rather, there is a process of many steps between the first contact of a sperm with an egg and the launching of the embryonic development of a new individual. In fact, it isn't until fifteen days after fertilization that the first structures destined to be part of the baby actually appear. What's happening during those first two weeks?

Picking up the themes of heredity and cells from Chapters 2 and 3, we delve into the way genes and cells work to create the multicellular structures that are your organs and, in the aggregate, your whole body. This is quite a trick, when you think about it. If all the cells in a body contain identical sets of chromosomes (it's true, as you'll learn), then how come a kidney cell is different from an eye cell? Why do we have all these different organs?

After we cover how embryonic development creates different specialized organs, we go right on to explain how all the major organ systems of the fully developed human body work.

CHAPTER 5: CYCLES OF MATTER AND ENERGY

Okay. Now that we understand how individual organisms are created and how they work, we back off a little and look at some general principles that govern the workings of all living things. And there is nothing more fundamental to the working of all life than the flow of matter and energy. The most basic point here is that all the energy used by living things (with one teensy-weensy exception) comes from the sun. Plants capture solar energy in the process of photosynthesis (at last, you're really going to understand how this works) and store it in the chemical bonds that hold together the atoms of molecules, such as sugar and other carbohydrates. Plants use that stuff to grow and animals eat plants, recycling the very same energy and the same molecules. Animals take the energy and use it to rearrange the atoms that make up plant molecules, changing the combinations to make the molecules of their own cells. And that includes us.

This is one of the reasons why it is not just an environmentalist cliché to say that all living things on Earth are intimately linked in a complex web of interrelationships. You could call them the food chains that bind. Which takes us inescapably to . . .

CHAPTER 6: THE INTERDEPENDENCE OF LIFE

Every living thing belongs to a community of other living things, and large groups of communities are linked into still larger ecological systems called biomes. We'll look at the many ways in which different species live together, why specific kinds of communities arise under certain environmental conditions, and how the various life forms interact. You'll learn the seven basic kinds of ecosystems and the four main kinds of relationships between species. (Remember symbiosis? We'll look at several different kinds, including some in which you, dear reader, are a participant.)

CHAPTER 7: EVOLUTION

If you've stuck with us this far, you may well be asking yourself how this marvelous, diverse—yet coordinated—system of genes and cells and bodies and ecosystems all came into being. The answer is that it wasn't easy and it took a terribly long time to happen. We'll go back to the mid-nineteenth century and sail around the world with Charles Darwin aboard H.M.S. *Beagle* and see what he saw—a vast body of evidence that led that young man to one of the most epochal discoveries in biology. There are three main bodies of evidence that lead inescapably to his theory of evolution, and we'll examine them all (actually, you'll already have learned them separately in the preceding six chapters) and see what they mean when taken together.

Then we'll get into Darwin's second major contribution: the mechanism that makes evolution happen—natural selection. You'll see that, contrary to the misunderstanding of some, it is anything but a totally random process. There are some very powerful forces in nature that make it happen.

Finally, after a quick tour of all the great epochs of Earth's ancient prehistory, we close with a bang—a look at the phenomenon of mass extinction and one very big hunk of rock that fell out of the sky millions of years ago (bigger even than any of those comet chunks that plowed into Jupiter). It was a bad day for the dinosaurs but, as you'll see, a boon for us mammals.

replied, "That the Almighty has an inordinate fondness for beetles."

Measured against insect diversity, the rest of the animal kingdom appears a piker with a mere 281,000 cataloged species—barely one-third of the whole. There are even fewer kinds of plants—275,000 known species. Bringing up the rear are three groups that are neither plant nor animal (we'll explain why later) and, therefore, merit their own kingdoms—the fungi (69,000), the one-celled protozoans (31,000), and the bacteria (5,000).

ESOTERIC TERMS
(es-ə-'ter-ik tərms)

Species: A unit of classification containing all the members of a population that are able to interbreed and produce fertile offspring under natural conditions.

Biodiversity: The total variety of different kinds of living things. In most biologists' minds, the more diversity in a given habitat, the healthier the habitat.

Taxonomy: The science of classifying organisms, not only by species, but by more inclusive categories, such as phylum, family, and order.

Vertebrates: Animals with backbones (which enclose the main nerve cord). There are five major groups of vertebrates: fish, amphibians, reptiles, birds, and mammals.

Invertebrates: All the animals that aren't vertebrates, which includes most species. They don't have bones. Major groups include worms, jellyfishes, insects, and spiders.

WE DON'T KNOW WHAT
WE DON'T KNOW

Despite centuries of collecting and cataloging, there are still lots of species out there that have not been counted, much less described and named. We know this because new species are still being discovered. Nobody breaks out the champagne, however, because it is a routine event and several new species turn up every day.

In fact, most natural history museums have growing backlogs of preserved specimens that field collectors have sent in but specialists have not had time to examine. Most are insects and other smallish beasts, but some are larger. In an average year two new bird species are discovered. For obvious reasons the oceans still harbor unknown beasts. Since 1908, for example, eleven new species of whale have been discovered.

Even new land mammals turn up. In 1988, a banner year, biologists learned of three new kinds of monkeys and monkeylike animals (in Madagascar and Central Africa) and a small deer, called a muntjac, that lives in the forests of China. Not all discoveries are in remote locations. In 1990, for example, a new monkey relative called the black-faced lion tamarin was discovered on an island just forty miles from the metropolis of Saõ Paulo, Brazil.

As biologist Edward O. Wilson has put it, "We dwell on a largely unexplored planet."

SO, WHO ELSE IS HERE?

How many species live on our planet? The short answer is that nobody knows. Experts agree that the true number has to be considerably greater than 1.4 million.

Some have guessed that there might be thirty million species, or, some say, even 100 million.

The thirty-million figure comes from a famous experiment in which insect researchers (entomologists who are human, that is, not six-legged scientists) hiked into the South American rainforest with a souped-up bug bomb. While standing on the ground, the scientific Orkin men blew a fog of fast-acting, quick-deteriorating insecticide high into one tree. It wafted higher than any human could climb—or swing a net. Within minutes they had to run for cover. A hail of insects, spiders, and other small stuff pelted sheets spread over the ground. Some were familiar species; many had never been seen before. The alien ants, beetles, flies, and others had spent their entire lives in a habitat that had never before been explored. Bombings of other tree species showed that each had its own community of never-before-seen critters. From the percentage of newly discovered insect species and the fact that there are 50,000 tropical tree species, the researchers calculated that if they were to bomb the whole rainforest, they would find about thirty million species.

But rainforest canopies are not the only unexplored regions on Earth. Aside from the dusty wastelands under beds, there is the ocean floor, from which dredge samples of bottom mud always turn up new species of worms, mollusks, and crustaceans.

And then there are the bacteria and other microbes. For each new species, it turns out, there is a whole community of microscopic organisms that live on it and it alone. Microbiologists (who, by the way, have normal stature) have cataloged fewer than 5,000 bacteria, but if you figure that each of the thirty million insects and

other life forms have just three or four microbial species
that live on them alone, the tally pushes past 100 million.

YOU ARE NOT ALONE

If you can't get too excited about critters you can't
see, there may be a couple of microspecies that *will* inter-
est you. They live, literally, in your face. Just inside the
follicles of your eyelashes (and everybody else's, probably) lives a species of mite—a wormlike animal with a
spiderlike head, barely visible to the naked eye. These
microbugs eat, live, mate, lay eggs, and just generally
hang out in that one environment. Don't expect a *National Geographic* special on "the magnificent eyelash
mite," but they're real and each of your eyelash follicles
is their wilderness habitat. What's more, there's a different mite that lives in the sebaceous glands of the fore-
head—those tiny pits in the skin that ooze oil and
torment teenagers.

Mites aren't just fond of humans. There's a species
that lives solely on the blood it sucks from the hind
feet of the soldier caste of a South American army ant.
Nobody's checked all the other species for similar cases.

Not only do we dwell on a largely unexplored planet,
each known and unknown body carries several largely
unexplored ecosystems.

CLASSIFICATION: THE GOOD, THE BAD, AND THE UGLY

Human beings have always looked at the many forms of
life around them and grouped them into categories. No

doubt one of the earliest and most important classification systems had just two categories: 1. edible, and 2. yuck! But after feasting on the first category, early humans had time to contemplate narrower distinctions. For example, some members of the edible kingdom could run away if you tried to catch them (animals) and some would stay put (plants).

Gradually people came to regard all snakes, for instance, as members of the same group, fish of another, antelopes of still another, and so on. But sometimes there was confusion. Birds and mice could be seen as belonging to different groups. But to which do bats belong? Are they furry birds or winged mice? Are whales air-breathing fish? Are humans naked apes?

The answers to such questions have come gradually and have depended on the interests of those doing the classifying. Most people are like the gauchos, the cowboys of Argentina famed for their horsemanship. Gauchos can classify horses into 200 categories according to their coloration but have just four categories for plants: fodder, bedding, wood, and everything else. Like gauchos, many Americans know a dozen or more categories for breeds of cats or dogs but aren't sure how cats are related to lions or dogs to wolves. Or, worse yet, how cats are related to dogs.

In the search for a more broadly useful system than gauchos and cat fanciers use, biologists have created an elaborate system of classification using a hierarchy of categories from the biggest—the animal kingdom, for example—to the smallest—the individual species. The original version was established in the 1700s by a Swedish naturalist named Carolus Linnaeus (see sidebar), who has become one of the immortals of biology.

The Linnaean system, as originally established, has seven levels: kingdom, phylum, class, order, family, genus, and species. The animal kingdom, for example, contains twenty phyla. Within each phylum are one or more classes, and so on. Sponges are one phylum, jellyfishes another. There are six phyla devoted to wormlike animals. Humans belong to one called chordata or, the chordates, which on first inspection is not a very exclusive club. Along with our proud kind, it includes creatures as different as fish, frogs, snakes, birds, elephants, and duckbilled platypuses.

No fair, you say? Worms get six phyla and we have to share our one phylum with all those other beasts? To be sure, there is some arbitrariness in the system, but that is

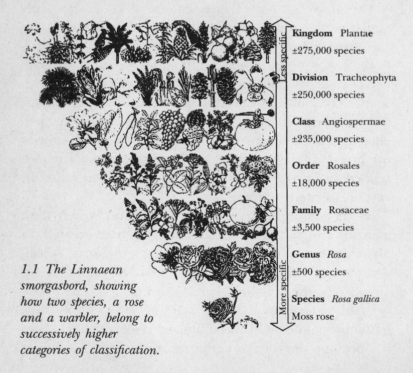

Kingdom Plantae
±275,000 species

Division Tracheophyta
±250,000 species

Class Angiospermae
±235,000 species

Order Rosales
±18,000 species

Family Rosaceae
±3,500 species

Genus *Rosa*
±500 species

Species *Rosa gallica*
Moss rose

Less specific

More specific

1.1 The Linnaean smorgasbord, showing how two species, a rose and a warbler, belong to successively higher categories of classification.

not the main reason. When biologists look at differences among species, they try to look past the superficial similarities and seek out fundamental resemblances and differences. All those worm-shaped critters may be long and cylindrical but inside, they possess some pretty radical differences, such as having a digestive tract or not. The feature that lumps our species with hagfish and armadillos is possession of a nerve cord that runs from head to tail. That means that, unlike all other animals, we chordates (except for a couple of very primitive types) have a highly organized nervous system under the control of one central processing center, a brain. That's a pretty significant criterion.

The Linnaean system has proven highly successful. It

Kingdom Animalia
> 1,000,000 species

Phylum Chordata
±40,000 species

Class Aves (birds)
8,600 species

Order Passeriformes
(songbirds)
6,160 species

Family Parulidae
(wood warblers)
125 species

Genus *Dendroica*
28 species

Species *Dendroica fusca*
Blackburnian warbler

Less specific

More specific

helps, for example, in figuring out whether to classify bats with birds. Both are chordates, but the bat's ability to gestate its young internally and feed them on mother's milk makes them mammals and definitely not birds, which lay eggs. Superficial structures like wings don't count at this level of classification. Thus, whales are mammals, too, and not fish.

Are humans naked apes? Almost. We are in the same kingdom, the same phylum, the same class, and even the same order. But that's where we part company. Apes and people are in separate families. Pretty close, but no banana.

In actual practice, it has turned out that some early categories were too narrow or too broad. So in order not to upset the whole system, biologists who classify organisms (who call themselves taxonomists, from the Greek *taxis*, meaning "arrangement" or "order") have created intermediate categories, such as suborders or superfamilies.

CLASSIFICATION: PERSON

Under the Linnaean system, human beings are classified as follows:

KINGDOM   Animalia (all animals)

PHYLUM   Chordata (animals with head-to-tail nerve cords and some primitive forms that have spinelike stiffening rods, called the notochord, but not nerves running along them)

SUBPHYLUM   Vertebrata (animals with nerve—spinal cord—encased in backbones)

CLASSIFICATION: PERSON (*continued*)

CLASS
Mammalia (all of the above plus having hair [at least somewhere on the body] and mammary glands to nurse their young)

SUBCLASS
Placentalia (all of the above plus a uterus in which the young grow before birth at a relatively advanced stage)

ORDER
Primates (all of the above plus retention of the primitive five fingers and toes, nails instead of claws, grasping hands, well-developed thumbs, stereoscopic vision, relatively large brain)

FAMILY
Hominidae (all of the above plus ability to walk on two legs and, some insist, to make tools. If you love Lucy—*Australopithecus afarensis*—you'll be glad to know she's in the family. Apes, however, are in the family Pongidae.)

GENUS
Homo (all of the above plus a mixture of traits that experts wrangle about incessantly, including larger brain size and head shape more like ours today than those of australopithecines)

SPECIES
Homo sapiens (all of the above plus having all anatomic features within the range of forms in living human beings. Or, if you prefer Mark Twain's criterion: *Homo sapiens* "is the only animal that blushes, or needs to.")

CLASSIFICATIONUS SYSTEMUS

Linnaeus invented not only the hierarchical system but the idea of designating each kind of living thing with a two-part name. The first, or genus, name is like a person's family name—Jones, say, or Simpson. The second, or species, name is like a person's first name—Bill, say, or Bart. Backwards, you may say, but it works. In the genus *Jones*, there might be two closely related species:

VIRUSES—UNWANTED, DEAD OR ALIVE

In the opinion of most biologists, viruses are not alive, so they aren't usually included in the classification system. But they are very much a product of life. They are, in fact, escaped genes—fragments of genetic code (in the form of DNA or RNA, which Chapter 2 is all about) that were once part of a truly living creature—that can weasel their way back into cells and force them to do their bidding.

What are a virus's orders? "Make more viruses that are exact copies of me!" Amazingly, infected cells obey and make so many more viruses that they break out and spread to other cells and other organisms.

So what is a virus? It is mainly a small piece of DNA or RNA wrapped in a protein coat. The protein happens to have a special shape, a microscopic "sheep's clothing" that makes viruses seem like something good. When the protein touches the right receptor on the cell's outer membrane, the cell mistakenly takes the virus in. Then the wolf—the virus's genes—come out. The cell can't tell the difference between virus genes and its own genes. So it obeys their selfish instructions.

Come to think of it, anything that diabolical must be alive and maybe even close to human.

Jones Bill and *Jones Mary*. In the genus *Homo*, there are three known species—*Homo habilis* (*habilis* is Latin for "able" or "adroit") and *Homo erectus* (*erectus* implying the ability to walk on the hind legs alone) and *Homo sapiens* (*sapiens* from the Latin for, get this!, "wise"). The first two *Homos* are extinct, the third is not yet. Incidentally, the convention is to capitalize the genus name but not the specific name.

The term genus can be confusing because if you use it by itself, it means genus, but if you use it with the second word, *sapiens* in this case, the two words together are the name of the species. In other words, human beings do not belong to the species *sapiens*. They belong to the species *Homo sapiens*. Also, note these words have tricky plurals. The plural of species is species. The plural of genus is genera.

Linnaeus started a good thing but no system is perfect. Sometimes after names have been assigned, subtle distinctions have been found. As a result, some species have been divided into subspecies. For example, organisms capable of reading this book are called *Homo sapiens sapiens* (an exaggeration, surely) and our close but vanished kinsfolk the Neanderthals are officially *Homo sapiens neanderthalensis* (the first skeletons found were in the valley, or *thal*, of Germany's Neander River).

LINNAEUS—LIVING SPECIES AND A DEAD LANGUAGE

The man who set up the modern system of classifying biodiversity wasn't really Linnaeus. He was a Swede named Carl von Linne (1707–1778) who liked Latin so

LINNAEUS—LIVING SPECIES
AND A DEAD LANGUAGE (*continued*)

much that he changed his name to Carolus Linnaeus. Though it sounds like one of his Latinized species names, it was actually just part of the same trend that caused a Pole named Kopernik to dub himself Copernicus.

In those days Latin was the language of scholars throughout Europe. By avoiding nationalistic sensitivities that might attach to the use of a "living" tongue, Latin provided a neutral means for scholarly communication. Nowadays, however, English has replaced Latin as the international language of science. Also, with the decline of classical scholarship, new species are now given all sorts of names, as often Greek as Latin, or sometimes mixing the two (which makes purists shudder). Or, very often, only thinly Latinized.

In setting up his system, Linnaeus took for himself the privilege of making up scientific names for the species he knew. It was he who chose the Latin *homo*, which means "human," presumably unaware it might someday be confused with the Greek *homo*, which means "same." Linnaeus did correctly classify whales as mammals and he recognized that humans were closely related to apes. But some would say he went too far when he named the orangutan *Homo troglodytes*.

Although Linnaeus's classification is recognized today as a remarkably accurate description of evolutionary relationships, Carl himself never caught on to what is perhaps the greatest fact in biology—evolution. (Much more about this in Chapter 7.) Linnaeus died a creationist thirty-one years before Charles Darwin was born.

NATURE'S REALPOLITIK: NOT TWO KINGDOMS BUT FIVE

When Linnaeus carved up the natural world, he decided there were only two kingdoms—plants and animals. In the two centuries since then, biologists decided that some species just didn't fit in. So most experts today accept a five-kingdom system: the familiar plants and animals plus a kingdom each for fungi, protozoans, and bacteria. Not bad for mold, slime, and germs. If you've never thought about how these organisms are different from plants and animals, you'll find their life-style as bizarre as anything a UFOlogist might dream up.

Here is how the living world breaks down according to that system, starting with the most primitive, the organisms that most closely resemble the first living things to appear on Earth.

KINGDOM SYSTEM

1. Monera—Bacteria and blue-green algae, which aren't plants but bacteria that can perform photosynthesis.

   Key features: One cell, but no nucleus, hence chromosomes sprawl loose all over cell. Eat all kingdoms.

2. Protista—Protozoans such as amoebas and paramecia and diverse others that used to be considered algae and fungi.

   Key features: Cells have a nucleus and other membrane-bounded organelles, making housekeeping much tidier. Many feed on bacteria.

KINGDOM SYSTEM (*continued*)

3. Fungi—Molds, mildews, rusts, smuts, truffles, and other yummy stuff.
   Key features: Usually many cells. Plantlike in some ways but unable to photosynthesize. Eat plants and animals, dead or alive.

4. Plantae—Plants, the most conspicuous form of life on Earth except maybe for Madonna.
   Key features: Many cells. Makes its own food by capturing solar energy through photosynthesis to make proteins, carbohydrates, and fats.

5. Animalia—Animals, from the sponges (not the blue ones in the kitchen) to humans.
   Key features: Many cells. Gets food by eating plants or animals that eat plants. Basically, it's an animal-eat-plant world.

KINGDOM MONERA

Named for the prosaic fact that they are single (mono) cells (even though another kingdom is also single celled), this is the realm of the bacteria and the blue-green algae (not to be confused with other things called algae that are really in two other kingdoms).

Bacteria are the oldest, smallest, and simplest living organisms known. (We're not counting viruses. See box, p. 12.) Bacteria are so small that five hundred lined up end to end would just about equal the thickness of a dime. Although many people, probably including your mother, think bacteria are harmful and a few are, the

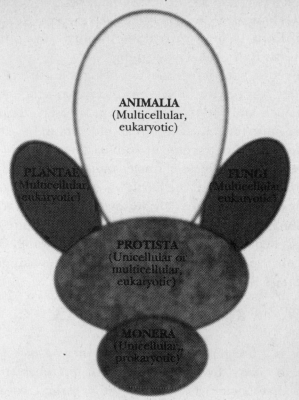

1.2 The family tree simplified, showing the five kingdoms that biologists now recognize.

vast majority are harmless and many are helpful. They turn milk into yogurt, gobble up oil spills at sea, and, along with Kingdom Fungi, they are the world's undertakers, recycling the corpses of other creatures.

The fundamental property setting bacteria apart is that their cells lack many of the internal structures possessed by the cells of all other living things. All other forms of life have a nucleus in each cell where the DNA, or genetic material, is kept. In bacteria, the genes simply float around loose inside the cell. Bacteria are the oldest known form of life on Earth and probably resemble the ancestors of all other living things.

The oldest known fossils of any living thing are of bacteria (they can't leave bones but their characteristic little ball-shaped and rod-shaped forms get preserved anyway) and are more than 3.5 billion years old. It must have been a snap for evolution to come up with bacteria, because these simple organisms (lacking nuclei and other internal structures) appeared when the Earth was only one billion years old. By contrast the next big advance in evolution took an additional two billion years. That was the invention of the single-celled organism with a nucleus and other organelles, the type of cell that reigns in all other kingdoms. (Chapter 3 is all about these cells.)

KINGDOM PROTISTA

This is a hodgepodge kingdom if there ever was one, home to a very diverse menagerie of microbes (though bigger than bacteria) and some rather large creatures as well. It includes the protozoans, one-celled organisms that often swim or crawl and which Linnaeus considered, with good reason, to be animals. In fact, *protozoa* means "first animals." These include such high school biology familiars as *Amoeba*, the amorphous blob that oozes over surfaces and engulfs its food, and *Paramecium*, the elegant, torpedo-shaped creature that hurtles through the water, propelled by a covering of hairlike cilia beating in waves. Maybe the best-named species in all of biology is the giant amoeba, terror of the microbial world, with the scientific name *Chaos chaos*.

In another part of this broad kingdom are the sea-

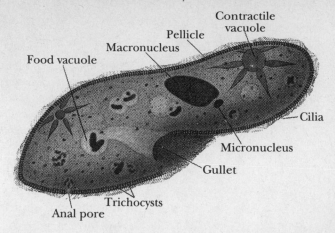

Food vacuole
Macronucleus
Pellicle
Contractile
vacuole
Cilia
Micronucleus
Gullet
Trichocysts
Anal pore

1.3 Paramecium, an organism so complex that it seems to have separate organs and tissues and yet it consists of just one cell.

weeds, ranging from one-celled algae to the giant kelps, which have lots of cells. Many of them carry out photosynthesis, like true plants, but they're still not plants because they lack the complex internal anatomy required for membership in that kingdom.

Among the most interesting creatures in the protist realm are those given the lovely name of slime molds. There are probably some in your backyard right now but, if not, they are definitely in the nearest wooded area. They have various ways of living but a typical slime mold is *Dictyostelium.*

During the happiest part of their life cycle, *Dictyostelium* exist as one-celled amoebas, slithering hither and yon over the dank surfaces of dead leaves and rotting vegetation, feasting on bacteria and other kibbles and microbits of organic matter. When the local climate starts to dry out and food is hard to find, however, the

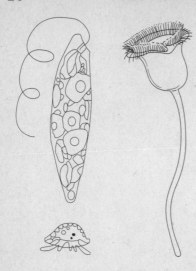

1.4 A Protist family album, showing some of the protozoans and one-celled algae (which aren't plants) that make up this highly diverse kingdom.

tough get going. They start converging on central locations, joining in crowded streams that look like webs of slime, usually on dead wood.

Then the amoebas assemble themselves into something that looks like a garden slug. Many one-celled individuals transform themselves into one many-celled individual. The slug creeps about for several hours and then, when it finds a place it likes, it settles down and changes form once again. It sends a thin stalk up into the air. Cells of the "body" then crawl up the stalk and form a ball, like one of those water towers that seem precariously balanced. As conditions turn still drier, the ball cracks open and dusty-dry cells (spores) from inside are expelled to the winds. They land and wait for wetter times. When that happens, they swell up and resume life as amoebas.

Just try accommodating that anatomy and life-style in any other kingdom.

It is from the protist kingdom that all "higher" forms of life evolved, namely the three kingdoms coming up next.

KINGDOM FUNGI

Linnaeus saw that mushrooms sprouted from the ground and didn't run away when approached. So he thought they were plants. But much later, plant specialists noted a very big difference. Fungi aren't green. They lack chloroplasts, the green things in plant cells that carry out photosynthesis, turning sunlight, water, and air into more plant stuff. (More on this in Chapter 5.) Instead, fungi send tentacles down into the ground or into dead trees to reach organic matter. The creeping fingertips ooze digestive juices and then soak up the goop. Only Sigourney Weaver could understand a creature like this. So fungi deserve a kingdom of their own.

Most of the "body" of many fungal species is a mass of filaments, penetrating the stuff it is feeding on. So you usually can't see it. It's underground or infiltrating a' rotting log. The parts we're most familiar with are, ahem, their reproductive organs—the parts we call mushrooms or toadstools. They poke up above ground, make spores (dried-up but viable cells), cast them to the winds to become new organisms, and vanish, often into the belly of one species of gourmet or another.

The fungal kingdom is also the source of other good things. Yeasts are essential to many forms of baked goods and cheeses, and to the fermentation of plant juices that yields beer, wine, and the stuff that is distilled into hard liquor. And, of course, the common bread mold, *Penicillium*, was the source of the world's first known antibiotic. The substance was, essentially, the mold's immune system, defending it against bacterial infection.

KINGDOM PLANTAE

They're so familiar, they hardly need an introduction. They're green, most have leaves (although some are in odd shapes like those of cactus needles), and most have roots that soak up water and minerals from the soil. And the green parts capture solar energy to make sugar. (Much more on this in Chapter 5.) In addition to the flowering plants, the kingdom also includes nonflowering species—mosses, ferns, and conifers like Christmas trees.

One major distinction of the plant realm is that all other forms of life (except certain bacteria and the communities that live on deep-sea, hot-water vents, to which we'll return in Chapter 5) depend on plants. All other kingdoms get their energy either by eating plants or eating animals that eat plants. That energy was originally the sun's. So, we're all solar powered.

KINGDOM ANIMALIA

At last, we come to our own realm. You might think you'd know an animal when you saw one and you'd be right most of the time if you said this or that *is* an animal. But you'd often be wrong when you said something was not an animal. For example, sea sponges are animals, as are corals (there are soft bodies living in the hard, stony stuff we usually think of as coral), sea fans, anemones, and a few other forms. Among the key distinguishing factors: Animals have many cells and get their food by eating other living (or not too dead) things.

The two main groups of animals are vertebrates (have bones) and invertebrates (have no bones). Animals have

1.5 Getting stoned. Chunks of coral rock secreted by tiny, soft animals called coral polyps.

complex sets of internal organs, specialized for sensing their environments, moving, feeding, digesting food, and reproducing. And they have nervous systems that communicate among all these organs.

SO, WHAT'S A SPECIES, ANYWAY?

Surprisingly enough, biologists are still wrangling over this one. The species is *the* fundamental unit of biodiversity, but if you check the books, you'll find that animals as different as a Chihuahua and a Great Dane belong to the same species while a leopard and a jaguar (which most people can't tell apart) are two different species.

Here's the simplest definition of a species: a population of organisms whose members can interbreed freely under natural conditions.

That holds for dogs of all breeds—all too well—and for

domestic cats. And it holds for the vast majority of animal and plant populations. One test of whether two individuals belong to the same species is whether their offspring are fertile. Horses and donkeys, two different species, ordinarily won't give one another the time of day in the wild, but with a little coaxing in the barnyard, they will mate and produce mules. But mules, though sturdy and stubborn, are sterile.

Consider this: Domestic dogs will interbreed with wolves and coyotes—and produce fertile offspring. Yet they are classified in three different species. They do, however, all belong to the genus *Canis*. Quite possibly biologists didn't know the three would interbreed when they gave them different species names. Maybe if they had it all to do over again, they'd just consider them different breeds of the same species.

Even harder problems arise when paleontologists try to classify extinct organisms by species. Take the evolutionary ancestors of human beings. There are four, maybe more, kinds of *Australopithecus* and at least three kinds of *Homo: habilis, erectus*, and *sapiens*. They're all dead, except for the last one, so we can't check on their sex lives. So, researchers go by anatomy. Basically, if two hominid skulls look sufficiently different, paleoanthropologists consider them different species. But this is such a subjective method that researchers actually refer to the "gestalt" of each skull when classifying it.

E PLURIBUS UNUM

For all the diversity of the living world, there is an underlying unity. All cells work pretty much the same way and the genetic code that governs all life forms is identical in all species. That's the whole basis of the biotechnology industry—that you can put a human gene in a bacterium and the bacterium can read it and follow its instructions.

The more you know about biology, the more amazing this is. Take the genetic code. Built into it is an arbitrariness; many other codes could work just as well. Why is there only one genetic code on Earth? Because, as we'll discuss further in Chapter 7, all life forms are descended from a common ancestor with that particular code. It is conceivable that early in Earth's history, there were several genetic codes and that, for some reason, all the others died out. Whatever the reason, all life on Earth is united in DNA. It is a bit divided in cellular structure and function. The bacteria have one kind of cell and everybody else has a different kind of cell.

PEEKING INTO ONE'S GENES

The old-time biologists—those practicing up until, maybe, twenty years ago—relied mainly on examination of anatomy to decide which species were which and how they were related. Nowadays you examine the genes. That's where genetic mutations happen (we'll discuss these agents of evolutionary change in Chapter 7), and each change leaves a record. The code is the same; it's just that there are, you might say, slight spelling differences. The human gene for the blood protein hemoglobin, for example, has several hundred letters that dictate the protein's structure. In other species, there are different letters at different positions. All the different spellings still make hemoglobin, but it has slight differences in structure from one species to the next—something like the way a Ford and a Chevy are different but still get the same job done. The more closely related two species are, the fewer the spelling differences.

The ability to examine genes has only come about in recent years, and it's still costly and time-consuming, but it is as close to a chronology of evolution as we may ever get.

By and large, the genetic comparisons that have been made so far confirm Linnaeus's classifications, but there have been some surprises. For example, humans turn out to be closer genetically to chimpanzees than gorillas are to chimpanzees. In other words, it may not make biological sense to classify humans in a group apart from the apes. Maybe Linnaeus wasn't so far off, after all.

SUMMARY

⏱ There are more than a million different known forms of life on Earth (defined as distinct species), but biologists are sure that the number of yet undiscovered species is several times higher.

⏱ Species are defined as populations of organisms whose members can interbreed freely under natural conditions and whose offspring are fertile.

⏱ Biologists classify species in larger groups—such as genera, families, phyla, and so on—according to whether or not they share basic (not superficial) features. Thus, for example, whales are classified as mammals (not fish), because the ability to breathe air and of mothers to nurse the young is seen as more important than living in water and swimming.

The largest divisions among life forms are at the kingdom level. The five kingdoms are, in common parlance: bacteria, protozoans, fungi, plants, and animals. The features that distinguish these five groups are the most fundamental—the most radical—that biologists can perceive.

Yet, for all the differences, there is an underlying unity. The genetic code is identical in all species and the fundamental metabolic processes within cells are the same. This is powerful evidence that all forms of life share a common ancestry.

CHAPTER 2

HEREDITY:
WHY HUMANS NEVER GIVE BIRTH TO WOMBATS

YOU MUST REMEMBER THIS

Genes govern almost everything in the bodies of living organisms but, like bureaucrats everywhere, they don't do anything unless they're told to. Nongenetic factors decide which genes are triggered into action.

Each gene's job is to tell a cell how to assemble raw materials into one particular kind of particle called a protein. Some proteins are structural building blocks of cells and tissues; others are enzymes that make certain chemical reactions happen.

THE LATEST FASHION IN GENES

If you want to keep up with the hottest news in biomedical science, you gotta know about genes. Every week or so they announce the discovery of a gene that is allegedly responsible for some disease or other condition. This week it might be the gene for zits, next week the gene for shopping. These may be important findings, but if you don't know what a gene is and how it works, you won't understand the true significance of the advances, which can be more—and, sometimes, less—than the news accounts suggest. For example, they could discover a gene for something and hold a press conference and get on *Nightline* but still not have the faintest idea what the gene actually does or how to use the information to help anybody. When you finish this chapter, you'll understand why that can be.

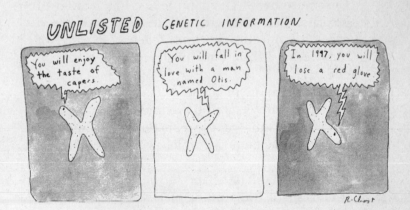

ESOTERIC TERMS
(es-ə-'ter-ik tərms)

Molecule: A particle made up of two or more atoms bound together. Bind two or more molecules together and it's still considered one molecule, only it's bigger now.

Protein: One of the fundamental kinds of molecules in living organisms. Some proteins are building blocks of cells. Others are enzymes, or catalysts, causing other molecules to undergo specific chemical reactions.

Gene: A unit of heredity that dictates the structure of one kind of protein molecule (both its composition and shape). The protein specified by one gene may be enough to determine a trait. More often genes work in groups to get a job done.

Chromosome: A chain of genes, like a train of boxcars. Each human cell (except red blood cells) has 46 chromosomes, each composed of hundreds or thousands of genes linked end to end.

DNA: The name of the chemical of which genes are made. Hence, also the name of the chemical of which chromosomes are made. DNA stands for *d*eoxyribo*n*ucleic *a*cid, which will not be on the quiz.

Double Helix: The shape of the DNA molecule. It consists of two strands of DNA, twined about one another like the handrails of a very long spiral staircase.

THE SOFTWARE OF LIFE

Unfortunately, the most common statement of what genes do—that they determine eye color, hair color, height, and other things inherited from parents—almost completely misses the point. The role of genes is far more profound. Genes also determine that a human being has two legs and can talk, that the head is on the upper end of the neck, and that red blood cells are inside the circulatory system.

Besides this, genes also govern such obscure events as the extraction of energy from food, the way brain cells store memories, and how well alcohol is detoxified in the liver. In other words, genes govern—or, at least, exert a powerful influence on—everything from the construction of molecules in cells to the architecture of the whole body and even aspects of its behavior.

Yet, genes do all this while remaining essentially passive, staying safely inside the cell's nucleus, sequestered from the biochemical hurly-burly just outside (the subject of Chapter 3). Like a computer program residing quietly on a disk but controlling a vast, automated automobile assembly line, genes just sit there and let other specialized molecules come along and, in a sense, read their messages. Toward the end of this chapter, we'll go into the details of this process.

So, what is a gene? It is a set of instructions, written in genetic code (we'll explain that in a minute) that tell a cell exactly how to make a given kind of protein—a wad of atoms that constitute one molecule of protein. Of course, the atoms have to be fitted together in just the right way for each type of protein. The gene tells the cell exactly how to do this.

Protein, by the way, isn't simply what you get in steak or soybeans. It is a certain type of molecule (more on this later, too) that comes in thousands of different forms, each with a specific job in the cell. The thingies inside muscles that create the force are proteins. The molecule in blood that carries oxygen from the lungs is a protein called hemoglobin. The enzymes in the stomach that digest food are proteins. The sex hormones, estrogen and testosterone, are proteins.

In virtually every cell of every living thing, there are thousands of different genes dictating the manufacture of thousands of different proteins. Protein molecules make up the largest part of the structures of our cells (which make up tissues which make up organs which make up bodies). And proteins serve as enzymes, hormones, and other agents of our survival.

It takes many different proteins working together to make one cell and to perform the chemical reactions that constitute life. The life of one microscopic cell may not be your idea of a good time, but put about sixty trillion of those little lives together and you have—you.

In each human cell, there are between 50,000 and 100,000 different genes. The number is the same in every cell of every person, but nobody knows what it is. All cells in an individual contain identical copies of the complete set of genes, but relatively few genes—perhaps between 1,000 and 5,000—are at work in any one cell. The others, like that ancient computer game that just sits on your hard disk, are dormant. Most of the active genes perform "housekeeping" chores, carrying out the metabolic reactions common to all cells. A few genes—perhaps a hundred that are active in any given cell—carry codes for proteins needed only in that type of cell. All cells, for example, contain a gene that guides the manufacture of the protein insulin (which the body needs to use sugar), but that gene is active only in certain cells of the pancreas. Pancreas cells make insulin, which the bloodstream carries to all other cells. In the same way, all cells carry the genes for making hair but, fortunately, they are dormant on, say, the tip of the nose. Also, for some unknown reason, the hair genes in the skin on top of some men's heads are switched off in middle age.

NUCLEAR ANNIHILATION

There is one exception to the rule that every cell has a full set of genes—the human red blood cell. Red cells are formed from the division of other cells in the bone marrow and, for unknown reasons, their nuclei (where the genes are housed) are pitched out of the cell and destroyed. Fortunately, this happens only after the cell has finished using the genes to guide the manufacture of a set of proteins.

This happens before the red cell enters the bloodstream. From then on, the red blood cell lives its 120-day life by recycling its original endowment of hemoglobin, the protein that carries oxygen, and other proteins. Hemoglobins that have ferried oxygen atoms from the lungs and released them, say, in the big toe are carried in red cells back to the lungs to get more oxygen.

GREG AND THE GARDEN PEAS: IT'S NOT IN THE BLOOD

One of the most important discoveries about how genes work was made long before anybody ever heard of genes.

Gregor Mendel (1822–1884), an Austrian monk, discovered in the 1860s that heredity is transmitted in discrete units and not in some method that blended the traits of Mom and Dad.

People had always known that children resemble their parents but until the twentieth century, most attributed this fact to a blending of bloods from the mother and the father. So common was the blood theory of inheritance that we still speak of inherited traits as "in the blood" or of "bloodlines." If Mom was short and Dad was tall, Junior usually turned out medium. It seemed obvious that inheritance was the result of blending.

The truth emerged in the middle of the nineteenth century, as a result of experiments Mendel did in his monastery garden. But Mendel didn't have a good press agent, and his epochal report in 1865 remained obscure until his writings were rediscovered in 1900. Mendel found that offspring were not a complete blend of Mom and Dad (you'd think that would have been obvious), but instead had inherited their traits in discrete and unchanging particles or units. Genes, we call them today.

Mendel worked on garden peas that happened to come in several pairs of traits. For example, some were tall, some short. They had either purple flowers or white ones. The peas were wrinkled or smooth. And so on. Mendel noticed that if he crossed opposites of a pair (by putting pollen from one on the flowers of the other), the progeny were never a blend. They always took after Mom or Dad, and sometimes the litter included a few of each.

Mendel kept detailed records of the outcomes. For example, every time he crossed a tall and a short pea plant (when both parents were like *their* parents), the progeny were always tall. That was a surprise. But if he crossed two of those tall progeny, he got a bigger surprise—only three out of every four offspring were tall. One was short, resembling neither of its tall parents but looking like one of its grandparents. A throwback.

After hundreds of such experiments, Mendel was not only sick of surprises, he got three ideas. The screwy results could be explained, he reasoned, if three things were true (we know now they are):

1. Each parent carries two units of heredity governing each trait. Today we call the units genes, which means that, like blue jeans, genes come in pairs.

2. When parents make their sex cells (pollen or sperm from the males, ova or eggs from the females), only one unit of heredity goes into each. In other words, each sperm or egg carries only one of the two units of a pair.

3. When the units combine in the offspring, the effect of one may dominate the effect of the other.

In the case of pea plant stature, for example, "tall" genes dominate "short" genes. (Without whips. This *is* sex but not the kinky stuff.) So if a plant inherits one tall gene and one short gene, there will be no blending. One tall gene, it turns out, supplies enough genetic instructions to make all the protein (whichever protein it is) needed to make the plant grow tall. The contribution of a second gene doesn't matter in this case. In genetic parlance that second gene is called "recessive." But if the pea inherited two short genes, there would be no tallness gene and the plant would have no choice but to be short, the default mode. Recessive traits prevail only when there is no dominant gene to oppose them.

The following chart shows how Mendel got the weird results he did when both parental peas were tall.

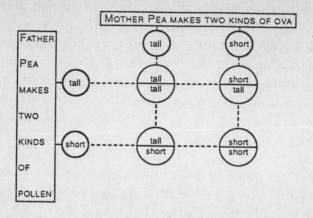

Three out of four of the offspring will be tall—one because it has only tall genes and two because they have one tall gene, which is enough since tall is dominant. The fourth offspring will be short because it inherited no tall genes.

By the way, "dominant" is not a value judgment. Shortness could just as easily have been dominant, depending on the function of the protein encoded by the genes in question. As it happens, six fingers in humans are dominant over five fingers. The only reason there are so few people with six fingers per hand is that practically everybody carries two copies of the recessive five-finger gene. A cleft in the chin, à la Kirk Douglas, is also the result of a dominant gene.

Mendel's pea-breeding experiments led to his epochal discovery that heredity was transmitted in discrete units. The notion of dominant and recessive genes, however, has turned out not to be as simple as the old pea breeder thought. This is because most traits (including such visible ones as height and skin color) are governed by several pairs of genes, not just one. So lots of different combinations are possible, sometimes creating situations that look, after all, very much like blending inheritance.

GREGOR MENDEL'S DATA FUDGE

Mendel is rightly credited with an epochal discovery but the good father may have been guilty of what today is called scientific misconduct. There is evidence he cooked his data.

Modern statisticians have gone back over Mendel's reported numbers of how many talls and shorts and so on he got in his experiments. The numbers he claimed are much closer to the statistical ideal than a real experiment would be likely to have produced. This would be akin, if Mendel had been counting the results of a coin toss, to claiming he flipped the coin ten times and got five heads and five tails. Try it a few times and you'll find the actual results are more likely not to be five and five but some other combination. Since coins have only two sides, we know what the "correct" answer should be.

Apparently Mendel had similar confidence in his new theory and felt he should "clean up" his data, maybe to make them look more convincing to an audience unfamiliar with statistics.

HOW WOULD YOU KNOW A GENE IF YOU SAW ONE?

Mendel showed that genes exist but as of 1900, when his long-neglected reports were rediscovered by three different biologists, nobody knew what they were made of, or where you might find them in the body. That would take another forty years. That didn't stop researchers from going ahead in 1909 and naming Mendel's units of heredity "genes," from the same root as "genesis."

Ironically, biologists of those days did know about chromosomes—the large structures that we know today are made up of lots of genes linked end to end. The word "chromosome" comes from the Greek for "colored body" because when scientists stained cells and looked at them under microscopes they saw these little colored bodies. So they called them colored bodies. But, being good scientists, they couldn't use ordinary words, so they turned to Greek. Actually, the colored bodies looked like little sausages. Maybe the scientists just didn't know the Greek for "little sausages."

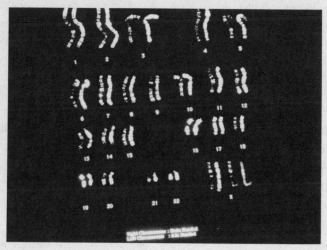

2.1 That's all she wrote. The entire set of genetic coding needed to construct a human being is contained in these 46 chromosomes. These have been photographed in a human cell under a microscope, the picture cut apart and chromosomes arranged in pairs by size and banding patterns. There are 22 pairs of equivalent chromosomes plus two sex chromosomes, in this case an X and a Y. It's a boy!

HOW GENES WORK
IN FOUR FAIRLY EASY LESSONS

1. How does the structure of DNA encode genes, which are the instructions for synthesizing protein molecules?

2. How does the cell make a working copy of the gene's instructions in the form of RNA?

3. How does the copy of the gene guide the manufacture of a protein molecule?

4. How does the protein fold into a useful form?

THE MOST FAMOUS SHAPE
IN BIOLOGY NEVER TO APPEAR
IN A CENTERFOLD
OR

1. How does the structure of DNA encode genes, which are the instructions for synthesizing protein molecules?

Before you can understand the workings of genes, you need to know a couple of things about the structure of DNA, the chemical of which genes are made. This is because the secret of life in the cell is in the structure of the molecules that make up DNA and all the other components of the cell. The shapes of molecules—along with the chemical nature of their constituent atoms—is what gives them their abilities. It's rather like a key or a wrench. If they aren't shaped right, they won't work.

So, first of all, DNA is a polymer. In other words, it is a long chain of smaller units linked end to end, like boxcars in a train. DNA's boxcars are called nucleotides.

Most of the time these DNA chains come in pairs— two DNA polymers twined around one another in the shape that James Watson and Francis Crick discovered

in 1953—the famous double helix. Helix is a fancy word
for the shape of a corkscrew.

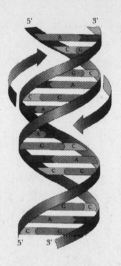

*2.2 This is what it's all
about—the famous
double helix showing the
bases labeled A, T, G
and C.*

OF SAUSAGES AND CORKSCREWS
. . . .

Don't be confused by the different shapes DNA can
take. The little sausages the old-time biologists saw
when they named chromosomes represent a form that
DNA takes only when it is preparing for cell division.
What the old-timers saw was DNA that had coiled and
shriveled into longish clumps.

All the rest of the time, DNA exists as sprawled out
threads, or strands, each a double helix. The DNA is so
sprawled out, in fact, that if you took the 46 strands
from a single human cell and lined them up end to end,
they would stretch an incredible three feet. Of course,
to fit inside a microscopic nucleus, the strands are very
narrow. Even the Japanese haven't topped that feat of
miniaturization.

You can visualize the double helix as a twisted rope ladder with wooden rungs. This is the form DNA takes when it is not doing anything. There are two times when DNA does something—duplicates itself for cell division and, the time we're interested in now, allows its genetic

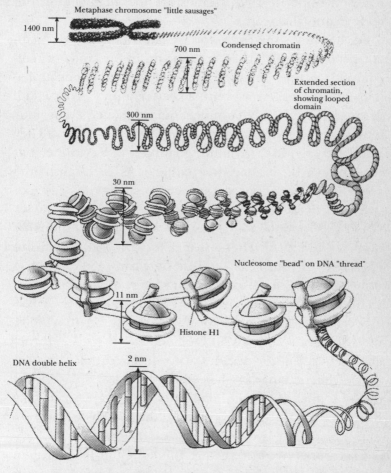

Metaphase chromosome "little sausages"

1400 nm

700 nm Condensed chromatin

Extended section of chromatin, showing looped domain

300 nm

30 nm

Nucleosome "bead" on DNA "thread"

11 nm

Histone H1

DNA double helix 2 nm

2.3 This is not a telephone cord. It is a chromosome in which a portion has been unwound, stretched out and magnified to show where the "double helix" fits in.

code to be read. On these occasions, the two strands of the double helix come apart. It is exactly as if someone sawed all the wooden rungs of the rope ladder down the middle.

The reason the two strands must come apart is that the genetic code is in the half-rungs. When the half-rungs of opposite DNA strands are locked together, the code cannot be read.

So, how does the code work? Glad you asked. It's actually pretty simple. There are four different kinds of half-rungs, each representing one letter of the genetic alphabet. If you learned a twenty-six-letter alphabet as a kid, you can manage DNA's much simpler system. The four letters of the genetic alphabet are A, T, G, and C. (They stand for adenine, thymine, guanine, and cytosine, but these won't be on the quiz either.) As in English, the sequence of letters on a DNA chain spells the message.

A genetic sentence (a gene) may start out TACGC-GAATTT and continue in that vein for hundreds or even thousands of letters. It may look like gibberish but it's poetry to your cells, because they can read it and use the coded messages to keep on living. Molecular biologists (biologists who specialize in the molecules of life, especially DNA) learned to read the code in the 1960s, unleashing a revolution in biomedical science that probably won't peak for decades. They learned that all words in the genetic language have three letters and each word in the sequence tells the cell what raw material to take first, second, third, and so on, adding each to the protein molecule it is making.

This works because of one fundamental fact: Protein molecules are polymers, too, chains of smaller units linked in exactly the same sequence as DNA's three-letter words. The subunits of proteins are amino acids and there are twenty kinds. The first word of the genetic sentence fragment in the previous paragraph, TAC, stands for the amino acid methionine. The next word (you'll notice genes don't have spaces between words) is GCG which, as every cell knows, is the amino acid arginine. Then comes AAT for leucine. And finally TTT, which is lysine. (Incidentally, molecular biologists don't

AMINO ACIDS: BODY-BUILDING BLOCKS

The protein you eat is made entirely of amino acids, but you can skip the Maalox. They're not very acidic. When you eat food protein, digestive enzymes (proteins themselves) break down the proteins into individual amino acids and small chains of a few amino acids. These are absorbed by the gut and shipped via the bloodstream to cells throughout the body for further digestion and recycling into new proteins.

The human body needs twenty kinds of amino acids to make all the 50,000 or so different kinds of proteins (one gene, one protein). Twelve of those amino acids are manufactured in the body, so it doesn't matter whether you eat them in food, although they are present in most foods. Eight of the twenty, however, cannot be made in the body. They must be consumed in food and are called the "eight essential amino acids."

Here are the names of the amino acids: alanine, phenylalanine, arginine, asparagine (named for asparagus; it's the aromatic stuff you smell in urine after eating the vegetable), aspartic acid (sold commercially as aspartame or Nutrasweet), cysteine, glycine, glutamine, glutamic acid (the stuff in MSG), histidine, leucine, isoleucine, lysine, methionine, proline, serine, threonine, tryptophan (abundant in milk, makes you drowsy), tyrosine, valine.

IT'S A DOGMA-EAT-DOGMA WORLD

Quite a few years ago, molecular biologists thought they had it all figured out. "DNA makes RNA makes protein," they declared and called it the "central dogma" of molecular biology. It was shorthand for "DNA guides the manufacture of RNA, and RNA, in turn, guides the manufacture of protein." The idea was that information (the genetic code) flows in one direction: to RNA and finally to protein.

Then they discovered an exception. Some viruses, it turned out, carry an enzyme that reverses the flow of information. The enzyme can take an RNA and transcribe its code back into DNA, literally creating a gene that is a copy of the RNA. They named the enzyme "reverse transcriptase." The AIDS-causing virus, HIV, is one of these and that's part of why it's so deadly. It actually copies its RNA genes into DNA and splices the DNA into the chromosomes of the cells it infects.

So the central dogma isn't so dogmatic anymore. But it is still a convenient way to remember the fundamental process of how genes work. DNA makes RNA makes protein. Say it over a few times and it'll keep the big picture clear as you continue this chapter.

always memorize the codes; they have cheat sheets pinned over their desks.)

IS IT ALIVE OR IS IT
MESSENGER RNA?
OR

2. How does the cell make a working copy of the gene's instructions in the form of RNA?

Before a gene can exert its influence on a cell (by causing a specific protein to be manufactured), it must be triggered into action. Genes, in other words, are lazy. They
don't do any work by themselves. In fact, they don't even
decide *when* to work. All they do is hang out in the nucleus,
waiting. They will not issue their instructions for protein
synthesis until a special molecule comes along with a shape
that can "recognize" the gene it is supposed to activate.
This molecule has a shape that fits precisely into a specific
sequence of nucleotides in the double helix. This regulatory molecule may, for example, be a hormone that has
just arrived in the cell from elsewhere in the body. It goes
into the nucleus and wanders around among the DNA
strands—something like a key trying out various locks—
until it gloms onto the correct DNA sequence. This sequence is not part of the gene itself but is an adjacent
stretch of DNA called the promoter region.

When the regulatory protein binds to the promoter
region, it changes shape slightly and becomes a perfect
docking site for a second kind of molecule to fit on—
the molecule that will carry out the process of making a
working copy of the gene. Working copies are necessary
because genes themselves are precious objects and must
be kept safely in the nucleus. Protein synthesis, on the
other hand, happens outside the nucleus. So, in effect,
a copy of the blueprints must be made and sent out to
the protein construction site.

The copying molecule is one of the two most amazing
molecular machines in biology. (The other one figures
in Part 3 of this process.) The molecular machine is
called RNA-polymerase—so called because it causes various loose building blocks of genetic code to link themselves into a chain (another polymer) very similar to

DNA but called RNA. RNA stands for *ribonucleic acid*, which is just like DNA in being made of four nucleotides except that instead of T, it has U, for uracil.

Here's what happens: First RNA-polymerase "unzips" the double helix. This is the same thing we did earlier when we sawed down through the wooden rungs. The two strands of DNA separate, letting the As, Ts, Gs, and Cs all hang out. Then the RNA-polymerase molecule proceeds to manufacture a strand of RNA that is an exact transcription of the DNA gene. The copy is called messenger RNA.

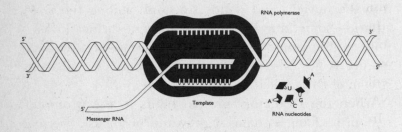

2.4 Unzipping the genes. Before a cell can read its genes and follow their instructions, it must unzip the double helix, exposing each strand of bases. Then a molecule called RNA polymerase transcribes the DNA into a new molecule called messenger RNA.

After unzipping the double helix (making, temporarily, two separate DNA strands), the RNA-polymerase creeps along one of the two strands (only one carries the code), moving like an inchworm on a twig. As it moves, one nucleotide at a time, it causes loose RNA nucleotides to link in a sequence that is the mirror image of the DNA sequence—a mirror image of the gene.

SHAKE HANDS WITH COMPLEMENTARITY

It's a concept that is absolutely fundamental both to genes and to hands. When you shake hands with somebody, the two of you clasp complementary hands—your right hand facing one way, the other person's right hand facing the opposite way. The shapes of the two hands, in other words, are complementary. The same principle applies when the DNA code of a gene is transcribed into messenger RNA. (And, as you'll see in the next chapter, it also applies when DNA duplicates itself for cell division.)

In both DNA and RNA, complementarity means that if one strand has a certain nucleotide at a certain position, the opposite strand will have a complementary but predictable nucleotide at the corresponding position. In DNA, the complements are always A with T, and G with C. In other words, A and T can shake hands, as can C and G. But no nucleotide can shake hands with a mem ber of the other pair. If we untwist the double helix for clarity, here is what complementarity looks like.

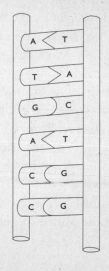

2.5 Thanks for the complement. Without it, we couldn't exist. This is a close-up of a piece of DNA showing how the four kinds of bases (a base is a rung in a nucleotide) can pair up only in the combinations of A-T or G-C.

In an RNA transcription of DNA, complementarity is the same except that where the DNA would put a T, RNA puts a U for the base called uracil.

The DNA can zip up behind the moving RNA-polymerase but there may be many molecules of RNA-polymerase working their way along the same gene, single file, each making one molecule of messenger RNA.

At the end of every gene is a stop sign, a code that tells RNA-polymerase to quit—ATT or ATC or, ironically, ACT. The RNA-polymerase falls off the DNA and releases the newly made messenger RNA, a faithful transcription of the gene's message.

JUST TELL ME THE GOOD PARTS

At this point, something weird happens. At several places along the length of RNA, there are stretches of pure genetic gibberish, segments of genetic code, faithfully copied from DNA, that even a cell cannot read. It would be comparable to this sentence *qwertyuiop* being interrupted *asdfghjkl* by jumbles of English letters that *zxcvbnm* make no sense. When molecular biologists discovered this a few years ago, they were flabbergasted but they have since come up with an intriguing theory that the gibberish plays useful roles in the evolution of new genetic messages. We'll get back to this in the last chapter, on evolution. Before the messenger RNA can direct protein synthesis, the nonsense segments are edited out. The job is done by—what else?—editor enzymes, special proteins that detect codes at the beginning and end of each gibberish segment, cut the RNA at those points, and then splice together the loose ends of the good parts.

LIFE ON THE GENE GANG
OR

3. How does the copy of the gene guide the manufacture of a protein molecule?

Okay, let's recap the play-by-play up to this point. Some kind of signal from the outside world arrived in the cell. Maybe it's puberty and this is a sex hormone trying to tell the zits to bloom. The signal molecule got inside the cell, traveled into the nucleus, and found the gene that it is shaped to recognize. When the signal glommed onto the gene, another molecule (RNA-polymerase, which is already in the nucleus) saw this signal and began making a working copy of the glommed-onto gene. RNA-polymerase manufactured a molecule of messenger RNA, an exact transcription of the gene. Then, because the first version of messenger RNA included a bunch of gibberish, another enzyme had to edit that stuff out.

Now we need to get the messenger RNA out to the factory floor, out to the place in the cell where proteins are synthesized. Actually, nobody knows how the messenger RNA gets out of the nucleus. Take our word for it, it just does.

Outside the nucleus the edited messenger RNA meets the second of the cell's most amazing machines—a gigantic molecule called a ribosome (RYE-bo-sohm). There are thousands, sometimes millions of ribosomes in a cell. Each is a complex molecule, made of nucleic acids (like those in RNA) and proteins. The ribosome reads RNA's message and assembles the specified protein. All proteins are, as the ''A'' students will remember, polymers— chains of linked subunits called amino acids. Like pop- beads of various colors and shapes, the twenty kinds of

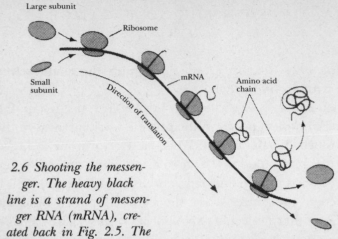

2.6 *Shooting the messen-*
ger. The heavy black
line is a strand of messen-
ger RNA (mRNA), cre-
ated back in Fig. 2.5. The
little ovals are the two parts of a ribosome, the machine that reads
the RNA and links amino acids in the sequence specified by
the mRNA. The protein is that little black line growing out of the
ribosome as it moves along the RNA from upper left to lower
right. As the amino acid chain grows longer, it curls and folds
into a characteristic shape.

amino acids can be linked in any sequence to become
any kind of protein. Typical proteins consist of scores or
hundreds of amino acids, linked one after the other.

CRACKING THE CODE

Now we come to one of the mysteries that baffled biologists for years. The genetic code is written with an alphabet of just four letters but there are twenty kinds of
amino acids. How can you specify twenty different things
when your code has only four different specifiers? Life
answered this question exactly as does English, which

uses a mere twenty-six letters to specify many thousands of different things. Like English, the genetic code does it with words—but genes talk only in three-letter words.

Ribosomes read messenger RNA in three-letter words. But it would be too simple to call them three-letter words, so molecular biologists say they are "triplet codons." For example, GAG (the RNA complement of DNA's CTC) is the codon for glutamic acid.

Ribosomes automatically attach themselves to messenger RNAs so that codons are brought into a special position one at a time. Drifting nearby in the cell is a stockpile of loose amino acids, each held in the grip of a short piece of another kind of RNA called "transfer RNA." Each transfer RNA includes its own codon but in a sequence complementary to the codon on the messenger RNA. We told you complementarity was important.

For example, the transfer RNA carrying one glutamic

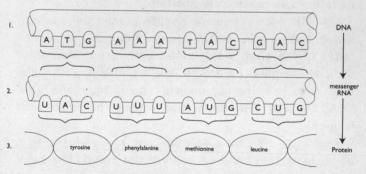

2.7 *Molecular biology up close and personal. This is a recap of the last few diagrams but magnified so you can see how the code works. At top is a single DNA strand, its bases looking remarkably like teeth. Below that is messenger RNA, made of three of the same bases but with U instead of T. It's coded sequence is the complement of DNA's bases directly above. Each group of three RNA bases is the code for one of 20 different kinds of amino acids, four of which are indicated here.*

acid molecule carries the codon: CUC. When CUC encounters GAG on messenger RNA, they fit perfectly and bind. The ribosome holds the amino acid in place, breaks off the transfer RNA that delivered it, casts off the now useless transfer RNA, and moves along the mes-

SICKLE CELL ANEMIA

One of the best understood genetic diseases is sickle cell anemia, which results from the tiniest possible spelling error in the gene for hemoglobin, the protein that carries oxygen in the blood. Actually there are two genes for hemoglobin. They dictate the structure of two proteins, designated alpha and beta. Every hemoglobin is made up of two alphas and two betas, which fit together in a four-leaf clover pattern. The code for alpha hemoglobin has 423 letters, specifying a sequence of 141 amino acids. The beta component has 438 letters, dictating 146 amino-acid positions.

The difference between a healthy person and one with sickle cell anemia, which can be fatal, is the second letter of the sixth word in the beta gene. Instead of CTC, which tells the cell to link a glutamic acid molecule next into the amino-acid chain, the gene says "CAC." As all cells know, that means the amino acid valine.

That little difference—a mutation, really—leads to a hemoglobin that behaves badly if the person has been working or playing strenuously. The oxygen deficit makes the hemoglobins link themselves into their own polymers, rigid rods that push out the red cell's membrane into bizarre shapes, including those of a crescent or sickle. Sickled cells jam up in narrow blood vessels, blocking blood flow and causing severe pain.

senger RNA to bring the next codon into position. When the next transfer RNA has locked onto the messenger RNA, the ribosome brings its attached amino acid into position to bind to the previous one.

Step by step, codon by codon, the ribosome creeps along the messenger RNA, parts of its structure swiveling and turning, flexing and extending like some wonderful contraption of levers and gears.

For all its chemical complexity, protein synthesis is astonishingly fast. Moreover, many ribosomes can work their way along the same messenger RNA simultaneously, all cranking out identical proteins. In a typical human cell thousands of ribosomes carry out more than a million amino-acid binding reactions every second, producing an estimated 2,000 new protein molecules per second. No wonder we're tired at the end of the day.

IT'S A WRAP!
OR

4. How does the protein fold into a useful form?

There's more to a protein than a chain of amino acids. Sorry about this, but there's a little more to explain. You can take some comfort in the fact that if you've gotten this far, you're nearly done with the hardest stuff in this book.

Once the protein emerges from the ribosome, it folds and curls into a shape determined by the chemical properties of each amino acid and its ability to interact with others in the chain. Parts of its length twist into helixes, other segments lock into rigid rods, still others act as swivels. One amino acid, cysteine, binds strongly to other cysteines, forming spot welds where the chain crosses itself.

Protein folding, as this process is called, is automatic,

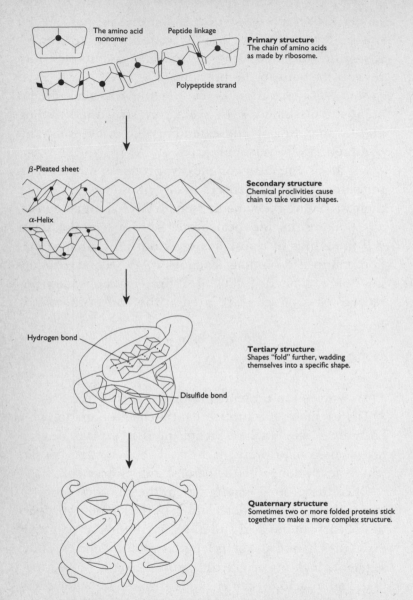

Primary structure
The chain of amino acids as made by ribosome.

The amino acid monomer

Peptide linkage

Polypeptide strand

β-Pleated sheet

α-Helix

Secondary structure
Chemical proclivities cause chain to take various shapes.

Hydrogen bond

Disulfide bond

Tertiary structure
Shapes "fold" further, wadding themselves into a specific shape.

Quaternary structure
Sometimes two or more folded proteins stick together to make a more complex structure.

2.8 Molecular origami. Once the chain of amino acids is formed, it starts pulling itself into folds and curls to take on its final, working shape.

dictated by the chemical proclivities of the amino acids. It is also consistent; a given amino-acid sequence will always, given the same chemical environment, fold up into the same shape. It is this shape and the chemical nature of the amino acids that wind up on the outside that give the protein its function.

Many proteins are further modified by the addition of sugars, fats, or other molecules that, essentially, dangle off the outer surface.

LANDMARKS
These Magic Moments

1865—Gregor Mendel (1822–1884)
Discovered that heredity is transmitted in discrete "units," later called genes. Asserted that each trait is governed by two genes, one inherited from each parent.

1902—William Sutton (1887–1916)
Recognized that chromosomes (whose function nobody knew) were like Mendel's "units." They come in pairs and newly formed organisms inherit one from each parent.

1920s—Thomas Hunt Morgan (1866–1945)
Found hereditary traits in fruit-fly breeding experiments were passed on in predictable groups. Proved genes were linked single file on chromosomes.

1930s to 1940s—Oswald Avery (1877–1955)
Showed that chromosomes were made of DNA.

LANDMARKS
These Magic Moments (*continued*)

1940s—George Beadle (1903–1989)
and **Edward Tatum** (1909–1975)
Established what genes actually do, i.e., each gene
dictates the structure of one kind of protein.

1953—James Watson (b. 1928)
and **Francis Crick** (b. 1916)
Showed DNA is a double helix. This revealed how
DNA can duplicate itself, each strand being a template for synthesis of a copy of the other.

1960s—Francis Crick (again)
and **Sidney Brenner** (b. 1927).
Figured out how DNA's paltry four "letters" could
spell all the messages genes need. Answer: the triplet codon. DNA uses three-letter words.

1966—Marshall Nirenberg (b. 1927)
and **Severo Ochoa** (1905–1993)
Cracked the genetic code, figuring out the DNA
"words" for each of the twenty amino acids.

SUMMARY

⏱ Genes do a lot more than determine eye color. They govern almost everything from the architecture of the body to the metabolism inside cells.

⏱ When active, a gene tells a cell how to make certain kinds of protein molecule. Some protein molecules act as the building blocks of cells and bodily tissues. Others are catalysts that carry out the chemical reactions inside cells that, in the aggregate, are called metabolism. Some, such as hormones, are released by one cell to travel as messengers to others.

⏱ The genes in human cells are packaged into 46 chromosomes, which are long strings of genes linked like boxcars. There are 22 pairs of nearly identical chromosomes plus two sex-determining chromosomes in a human cell. At conception Mom's egg brings 23 chromosomes (one member of each pair) and Dad's sperm brings 23.

⏱ DNA, the chemical of which genes are made, is a chain of smaller molecules called nucleotides linked like beads on a chain. There are four different kinds of beads and, like letters of the English alphabet, they can be linked in any sequence.

⏱ When the gene receives the right chemical signal, it swings into action and lets a special enzyme read its code. The enzyme copies the code into· a new kind of molecule, a long DNA-like chain called messenger RNA. This molecule then moves to the cell's protein-making factory, called a ribosome. The ribosome reads the code carried by the messenger RNA, grabs free-floating amino acids and links them in the specified sequence. Then, the chain of amino acids folds itself into a compact blob that is, essentially, its working form.

CHAPTER 3

CELLS:
THEY'RE TINY BUT—
STAND BACK!—
THEY'RE *ALIVE*

YOU MUST REMEMBER THIS

The unit of life is the cell and cells of all species are all pretty much alike in the way they are built and the way they work. Some creatures consist of just one cell, while others—such as the organisms reading this book—are made up of trillions, all living together in giant, walking colonies. Wait 'til Steven Spielberg hears about this.

AND YOU CAN FORGET THIS

Protoplasm. Despite what they told you in school, there is nothing inside cells that goes by this name. Instead, cells are crammed with specialized chambers full of molecular machinery, pathways linking the chambers and parcels of cargo being shuttled from one place to another for processing. Calling it all protoplasm makes about as much sense as calling what's under the hood of a car "mechanical matter."

It is the working together of the cell's internal machinery—all strictly according to the laws of chemistry and physics—that constitutes life. There's nothing miraculous about the "miracle of life."

CREEPY AND CRAWLY

Please don't be alarmed, but there are several billion tiny creatures crawling about inside your body. Right now. Some are slithering around in your lungs, eating dust particles you inhaled. Others are cruising your veins, hunting down and gobbling up dead blood cells. Still others roam the tissues of your body, looking for cells that have turned cancerous. These microscopic beasts, each looking and acting rather like a one-celled amoeba, flourish throughout your body, pursuing their grisly little lives, day in and day out.

Like it or not, these critters are part of you. They are cells that belong to one of the most amazing and tightly knit communities imaginable—the aggregation of about sixty trillion cells (and stuff made by those cells, such as bone) that make up each adult human being.

Fortunately, most of the other cells don't crawl

COMPARATIVE BIOCHEMISTRY

3.1 Bagels assaulting Mt. Everest? Actually, it's red blood cells on the point of a pin.

around; they are content to stay put, which means your liver won't go for a hike and the thigh bone will still be connected to the knee bone. While most cells are stay-at-homes, each is still a living organism in its own right and every human body is the sum of all those many parts.

Human cells vary greatly in size, from the tiny red blood cell at 1/25,000 of an inch (0.00004 inch) across to a typical diameter ten times larger—1/2,500 of an inch (0.0004 inch)—for a kidney cell or a liver cell to the gigantic muscle cells that can be thin filaments a few inches long to certain nerve cells that can grow four feet long. (The record holders are the nerve cells that begin at the base of the spine and run all the way to the tip of the big toe.)

ESOTERIC TERMS
(es-ə-'ter-ik tərms)

Cell: The smallest unit of matter that can be considered alive.

Life: An aggregation of molecules organized in such a way that, when sufficiently wet, it can take in nutrients and use the matter and energy to grow and reproduce itself. The cell is the smallest thing that can do this.

Organelles: The "organs" of cells These are specialized internal compartments or molecular structures that host specific groups of chemical reactions. Nearly all the reactions require energy.

Protein: A type of molecule that can be made in thousands of different forms, each capable of performing a specific job. Some are structural building blocks of cells. Others are enzymes, or catalysts—molecules that cause other molecules to change in specific ways. Still other proteins are information carriers within the body. The structure of each protein is dictated by a gene.

Gene: A segment of a chromosome (made of the chemical DNA) that dictates the structure and composition of one kind of protein. Cells use the guidance of genes to control the synthesis of proteins.

There are about two hundred different kinds of cells in the human body with different shapes and jobs, but all are similar in basic structure and internal workings.

LIFE IN THE DISH

The cells that live obediently as members of the colony we call the body are only too happy to escape if they can find a friend on the outside. Snip off a few cells of skin, say, or kidney or heart (don't try this at home, folks) and put them in a dish of water with dissolved nutrients. The cells will immediately revert to the lifestyle of their most ancient evolutionary ancestors—crawling around the bottom of their dish as if they were microorganisms on the bottom of a pond. They'll feed, grow, and even reproduce by cell division.

As the cells creep about, they thrust out a wide, ruffled edge that seems to search the space ahead. Parts of the leading edge stick down to the surface and the cell's body appears to drag its hindquarters along. When two cells bump, they cringe, shrink back, and quietly slither off in new directions.

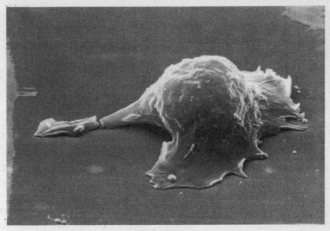

3.2 A rat cell crawling on the bottom of a culture dish.

LIFE IN THE DISH (*continued*)

Cell biologists have been growing cells in dishes for most of this century and using those cells as tiny experimental animals. The cells must be fed periodically and their numbers thinned out when they get too crowded.

Most cell cultures taken from normal body tissues will grow and reproduce for only a few generations before dying. But cells taken from cancers are potentially immortal. The very property that made them deadly to their former hosts now makes them invaluable to biomedical researchers.

Perhaps the most widely known cell culture contains what are called HeLa cells, named for a Baltimore woman, Henrietta Lacks. She died in 1951 of cancer, but her tumor's cells are growing today in hundreds of laboratories all over the world. In fact, the total weight of HeLa cells alive today is far greater than Henrietta's whole self ever was.

AH, SWEET MYSTERY OF CELL THEORY

Until the 1800s nobody understood about cells. People thought that life came in units of whole organisms—the rutabaga, say, or the raccoon. They knew there were various squishy objects inside organisms but were content to think that these (organs, we now call them) were simply different arrangements of fibers and tubes and that these just, somehow—don't ask so many questions—grew, maybe like crystals.

Nine out of ten doctors agreed that a mystical, unknowable "vital force" acted inside all living things—a kind of magic "breath of life" that some equated with the soul. The theory was known as vitalism. As long as the force was with you, you stayed alive.

Then in the 1830s, two German biologists, Matthias Schleiden and Theodor Schwann, advanced a schwell idea that came to be called "cell theory." They had been looking at thin slices of animals and plants under microscopes and they saw something odd. Everything seemed to be made of lots of tiny parcels or chambers, just as a brick wall is made of lots of bricks.

3.3 Is facial hair correlated with scientific achievement? Matthias Schleiden, right (1804–1881), co-founder with Theodore Schwann, left (1810–1882), of the cell theory, which held that organisms are made of discrete units of life and called cells.

They called the chambers cells, resurrecting a term that an Englishman named Robert Hooke had used two centuries earlier when looking at a slice of cork (tree bark that once was alive) using the world's first compound microscope, which he just happened to have invented. Hooke saw rows of rectangular holes that reminded him of monks' cells in a monastery. So he called the spaces cells. But that's as far as Hooke went. There was nobody home in the dried-out cork cells.

Schleiden and Schwann revived Hooke's term and applied it not to the empty spaces but to the stuff inside those spaces. S&S went further. They said cells were complete packages of life—the smallest clumps of matter that could carry out all the phenomena that constitute life.

THE MIRACLE OF THE DUNG HEAP

Long before the mechanists rose to power in biology, people used to think that life could emerge spontaneously out of practically anything, especially disgusting stuff. Heaps of manure seemed to give rise to worms and rotting garbage, as anybody could see, caused maggots to come into existence.

The first refutation came from Francesco Redi, an Italian scientist working in the 1600s, who did one of science's first controlled experiments. He put some meat in two jars, covered the top of one with gauze, and let the meat rot. Maggots appeared in the uncovered jar but not in the other. Redi concluded that if flies couldn't get at the putrid stuff, they couldn't lay eggs in it, and, amazingly enough, no eggs could hatch into maggots.

In the 1700s, another Italian (they dominated science before the Germans gained sway) named Lazzaro Spallanzani took the case against spontaneous generation further. Despite Redi's proof, people just couldn't shake the notion of continuing miracles of the creation of life. They still thought microorganisms popped into existence in stagnant pools of water. He did an experiment similar to Redi's except that one jar of water was designed so that bacterial spores (which drift in air) could not fall into it. That jar stayed clear while another (into which air currents could carry spores) turned cloudy and slimy with tiny critters.

Cell theory was the first solid blow to vitalism. Instead of a magic force that accounts for life, cell theory suggested life was a result of entirely natural processes, phenomena that could be studied objectively. Schleiden even went so far as to pose one of cell theory's most fascinating philosophical questions: If a cell is a living thing, then is a multicellular organism really a highly cooperative colony of one-celled individuals? This proposition raises the deeper question of why a group of presumably sensible cells would cooperate to form, say, Howard Stern.

For all his brilliance, however, Schwann could not completely shake the old vitalist belief in mystical processes. He claimed that cells arose spontaneously, materializing out of some kind of strange bodily fluid. It would take another German (they came to dominate science in the 1800s) to make the next great advance—a pathologist named Rudolf Virchow.

3.4 Rudolf Virchow (1821–1902) modified cell theory to say that cells arise from the division of other cells.

In 1850, Virchow ran with cell theory, pretty much creating the modern view of the nature of cells. He said all cells came not from any precious bodily fluid but from other cells that had split in two. In his microscope he had seen cells pinching in two, and each of the two "daughter cells" (for obscure reasons, it's always daughter cells, even in sons) going on to live a life of its own, eating and growing big enough to split again.

At last, science had come up with a reasonable way for a creature to grow (repeated splitting and growing of individual cells) and for it to make babies (shedding cells—more fun than it sounds—that can start the process all over again). Life, in other words, is a property of cells and is passed on by them to new generations of cells (and new organisms) in an unbroken chain.

Virchow did even more. He said all diseases were the result of problems arising within cells. He was right, and this is why the scientific discipline of cell biology (which includes molecular biology) is the frontier of almost all biomedical research.

There was one more aspect of Virchow's contribution. He said life was a natural phenomenon. In the nineteenth century that was saying a lot, because it meant he was dismissing any role for supernatural phenomena. Vitalism, he declared, was wrong. Instead he said all the internal workings of cells could be explained entirely as mechanical processes, as natural chemical reactions.

Virchow and his followers came to be called mechanists and they would do battle with the vitalists for another seventy-five or so years, advancing their views in provocative terms. "A man is what he eats," one slogan put it, noting that every atom of the human body has been extracted from food. (Of course, to be totally accurate, the mechanists should have included water and air, both of which contribute atoms.) Not even thinking, they argued, has a supernatural foundation. "The brain," the mechanists liked to say, "secretes thought as a kidney secretes urine."

In the years since, the mechanist view has come to dominate biology totally. About the only place you still hear tinges of vitalism is among those who study the human mind. The simple fact that consciousness exists (at least we hope you're still awake) reveals to us such an astonishing phenomenon that it is hard to imagine how it can be accounted for by the mere shuttling of electrochemical impulses among neurons. (See Chapter 4's section on the nervous system.) How can a lump of meat (such as the brain) be conscious? The short answer is that nobody has the faintest idea. And so, in this one area, vitalism still serves its ancient role: an easy explanation for what we cannot yet explain.

Except for that, the mechanist view rules. In fact, many of the most intimate processes within cells are now so well understood that they can be duplicated in the test tube. Even genes, for example, can be manufactured using off-the-shelf chemicals and then made to duplicate themselves, exactly as if they were in a dividing cell. These synthetic genes can be put inside a cell (injected with a syringe and a very fine point), and the cell will treat them exactly like a natural gene, following their instructions to the letter.

Cell biologists can't yet explain every detail of how life works, but in recent years laboratories have made dramatic advances. They can now explain many of life's most intimate workings as the results of purely nonliving events—interactions among molecules or atoms no more mysterious, though usually far more complex and wondrous, than the crystallization of water molecules into snowflakes.

The phenomenon of "self-assembly" explains many events in cells that once seemed utterly mysterious. It is akin to learning that the steel skeleton of a building will assemble spontaneously once a load of girders is dumped at a construction site. The cell's genes cause a load of pro-

SOME ASSEMBLY REQUIRED

One of the most fundamental concepts needed to understand how life works is the phenomenon called "self-assembly." The basic point is simple. Molecules have built-in propensities to combine with other molecules in predictable ways. Move two magnets close together on a tabletop and they suddenly stick together. There's nothing magic about it; that's just what magnets do. Most of the molecules of life are the same way. Like magnets they may be attractive or repulsive. In addition, they have other ways of combining and, as everyone should be, are very choosy about what they will combine with.

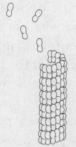

3.5 Peanuts with a mind of their own. This is a microtubule, one kind of "skeleton" inside cells. Tubulins, like many other proteins, spontaneously assemble themselves into more complex structures and automatically link up into long hollow tubes.

tein molecules (the building blocks of many structures inside cells) to be synthesized, but the proteins, needing no further control, spontaneously assemble themselves into larger structures. The final result, in other words, is dictated by the structure of the components. There is only one way the components can fit together. And the components spontaneously fit themselves together. Hard as it may seem to believe, the girders leap into position and rivet themselves in place. (Actually, they don't "know" what position to move to. They simply bounce around and fail to stick in any position but the right one.)

SUSPENDED ANIMATION

After the mechanists laid to rest the notion that life could arise spontaneously, there remained one niggling observation that was hard to explain. There are certain creatures that can dry out and shrivel up—literally turn crisp and dry as dust and stay that way for years—kinda like upstairs at the Bates Motel. But when put in water, the animals "come back to life."

The species that can do this are small soil and pond dwellers such as nematodes, rotifers, and tardigrades.

For generations, biologists debated whether the animals died and came back to life. One side held that dried was dead and called themselves "resurrectionists." The other side said that amounted to spontaneous generation and argued that the animals might look dead but were still carrying on a very slow rate of metabolism. This side, of course, called itself "anti-resurrectionists."

Only in the last twenty years have researchers settled the question: Neither side was exactly right.

It turns out that all metabolism truly stops. What saves the animals is that when they are drying out, their cells make special compounds that protect the cells' innards. When the dried cells soak up water, they simply resume metabolism because all the necessary machinery is in place.

LIFE IN THE FAST LANE

If you take a single cell and put it under a powerful microscope, you'll see something that might remind you of an aerial view of a city—innumerable objects zooming around, bumping and jostling in a manner reminiscent of the view of city streets from high atop a skyscraper. Like city streets jammed with cars, people, and buses, the cell's innards are crowded with objects of many shapes and sizes, many of them moving around. The image is

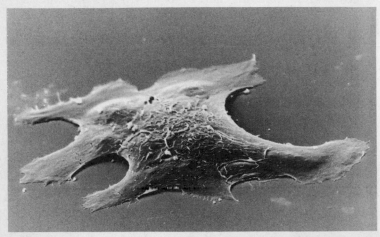

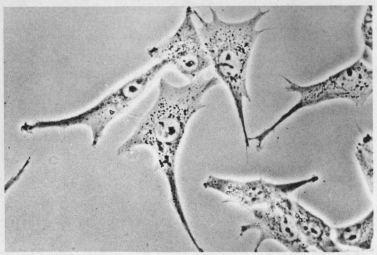

nothing like the typical textbook diagram of a cell, which shows a few lonely objects drifting in an empty intracellular sea. In a real cell there is no vacant space.

Under the microscope, you'd see thousands of tiny

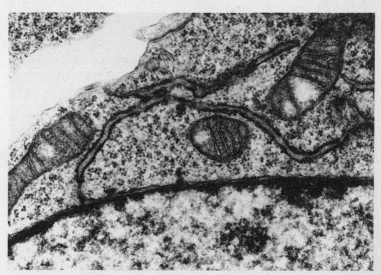

3.6, 3.7, 3.8 Three faces of the cell, each taken by one of the three major types of microscope used to study these beasts. If you simply ran into one on the street and could see only its outer surface, it would look like a cell crawling over the flat bottom of a culture dish. This image (facing page, top) is made with a "scanning electron microscope." If you used a typical light microscope, you could see inside the nearly transparent cell, but only a few of the organelles would be visible. Those big ovals (facing page, bottom) are the nuclei. To go up close and personal, you need a "transmission electron microscope," which took the third picture (above) of a thin slice through a cell. With this you can even peer inside the mitochondria (the stripey-looking things). This is such a close-up view that you can only see one edge of the nucleus (at bottom).

spheres, each a hollow container of chemicals, jostling about. Some jump crazily. Some glide in straight lines—some smoothly, some in fits and starts. Dark, sausage-shaped objects loom into sight, turn a corner, and slither back down out of the microscope's focal plane. Not only are there cells crawling around inside you, there's stuff crawling around inside cells. That's life—dynamic even at its most fundamental level, a ceaseless concert of motions and of thousands of simultaneous chemical and physical reactions.

But life is not the product of random reactions in some kind of biological broth inside cells. It is a phenomenon that emerges from the cell for two nonrandom reasons, two very important properties that keep every cell highly organized, which is to say, alive.

First, cells (of all species except bacteria) have internal compartments, each containing different sets of chemicals. These chambers, made from a membrane similar to the one that encloses the whole cell, are like the bottles and test tubes of a chemistry set. Each kind of chamber has its own standard set of chemicals and can perform certain chemical reactions on the materials that come into the chamber. Most of the bubbles and blobs that jiggle and lurch under the microscope are the cell's "organelles," which simply means "little organs." (We weren't kidding when we said cells are critters in their own right.) Like organs in the body, organelles come in various specific shapes and sizes, each with a particular job to do.

The second form of organization is a transportation network that links the bubbles and blobs, ferrying the products of chemical reactions from one compartment to another. At any moment, thousands of tiny, chemical-laden, membrane-enclosed bubbles, called vesicles, are being transported from place to place in the cell. And when we say "transported," we mean it. Cells have lots of tiny molecule-sized motors that can hook onto a vesi-cle and physically drag it to another place.

Cells do this not just to pass the time but because they need to move substances that are made in one place to some other place where they are processed (changed in some chemical way) and then to still another place to be used.

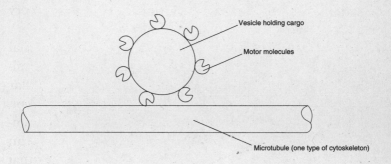

3.9 *The "Teamsters" of life, you might call the motor molecules that haul cargo throughout the cell.*

How do the motors know where to go? Simple. They follow tracks, rather like railroad tracks, that lead from one organelle to another. The motor doesn't have to see or steer. It simply pulls itself along the track like a locomotive, going wherever the track goes—while staying hitched to its vesicle cargo.

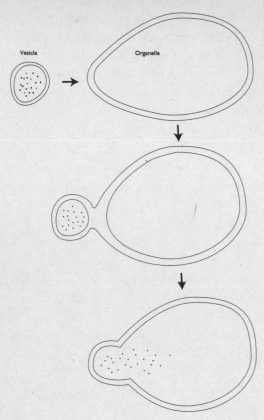

3.10 Special delivery. When the cargo reaches its destination, the vesicle's membrane fuses with that of the receiving organelle. The process automatically dumps the vesicle's contents into the organelle.

Vesicles even carry address labels. Sort of. These labels are specific kinds of molecules that stick out of the vesicle. As the vesicle tootles around the cell, it bumps into other vesicles and organelles. Nothing happens unless the address label molecule encounters an organelle carrying a complementary molecule on its outer surface. If

so, the two molecules fit together and the two compartments merge. In effect, the little one dumps its cargo
into the big one. The vesicle is like a drunk with a key
to his own house but no ability to figure out where the
house is. He simply goes down the streets, trying every
lock until one opens.

Keep the lock-and-key metaphor in mind. It's fundamental to lots of different things that happen in cells.
Usually the molecules that serve as locks and keys are
proteins—the multipurpose, workhorse molecules that
come in thousands of different but specific forms, each
prescribed by a gene. Proteins that fit together this way
make many different kinds of things happen. As one
example, it's how viruses infect cells. It's how sperm get
inside eggs. Cell biologists often call the lock a receptor.
Certain cells of the immune system have receptors for
the AIDS-causing virus—not that that's why the receptor
is there in the first place. It's just that the virus happens
to have a protein on its surface that resembles a different
one that the receptor is supposed to accept. Once a receptor receives its intended partner, it does something.
Some receptors pull the partner through the membrane
in which they're setting. Other receptors leave the partner outside but respond by using their tails, which are
inside the cell, to activate some other process.

HEY, KIDS! BUILD YOUR OWN CELL! HERE'S WHAT YOU'LL NEED

Remember the protoplasm we told you to forget about
at the beginning of this chapter? If you followed direc-

tions, that's okay. We'll remind you. Protoplasm is what biologists in the old days said cells were made of. They said it was some kind of gelatinous stuff. They said that because they didn't know any better. Today, biologists have looked a lot closer and discovered that the jellylike stuff is made of lots of discrete objects of several different kinds. Each has a specific job to do in the cell.

Biologists don't call them "objects." They call them "organelles." So, if Revell ever comes out with a cell kit that you can put together at home, here are the organelles it's going to contain.

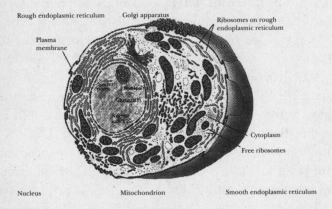

3.11 Your average animal cell (plants are slightly different, but we're not going to worry about that).

THE NUCLEUS—MISSION CONTROL

Almost every cell has one of these. The nucleus is, by far, the biggest organelle, occupying about one-tenth of the cell's volume. It's the control center—the cell's brain—and, basi-

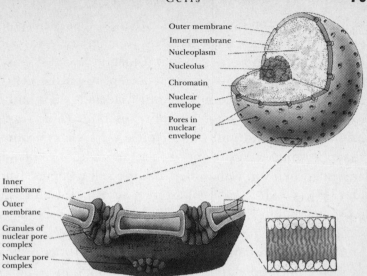

Outer membrane

Inner membrane

Nucleoplasm

Nucleolus

Chromatin

Nuclear envelope

Pores in nuclear envelope

Inner membrane

Outer membrane

Granules of nuclear pore complex

Nuclear pore complex

3.12 Mission control for cells is the nucleus, which has an un-usual double-walled membrane called a nuclear envelope. Note the nuclear pore complexes, which are special ports controlling entry and exit of molecules. The nucleolus makes ribosomes. "Chromatin" is a name for unraveled chromosomes.

cally, runs the show. That's because it contains the main re-pository of genes. As we learned in Chapter 2, each of the thousands of genes in a cell tells it how to make one particu-lar kind of protein. Those proteins, in turn, carry out the chemical reactions that are life. Among the proteins, for ex-ample, are the enzymes that digest food in the gut, the mole-cules in the back of the eye that create an electrical signal when light strikes them, the fibers in muscles that contract to make the muscle work, and thousands of others. Each of these proteins has a specific chemical composition and over-all shape that is dictated by one gene.

In human cells the genes are linked end to end in 46

strings—46 chromosomes. The chemical that genes and chromosomes are made of is DNA. Remember, as we said before, the chromosomes are not usually in the fat X shape most often depicted in pictures. Except when the cell is dividing, the chromosomes are unraveled and exist in the form of long threads—*very* long threads. If you could reach into a nucleus and pull out a typical unraveled chromosome, it would measure more than an inch long. This is amazing when you remember that the cell it comes from may be less than 1/1,000 of an inch across. Link all chromosomes end to end and they'll extend three feet.

Enclosing the nucleus is a special double-walled membrane (it's important to keep the DNA safe), perforated by lots of small openings. Coming in through these pores are molecules that will search out specific genes and switch them on, starting the process in which the gene's coded message will be transcribed into a similar long-chain molecule called messenger RNA. Going out are messenger RNA molecules (Chapter 2), carrying a transcribed copy of the gene's instructions to the cell's protein factory, the ribosome.

THE RIBOSOME—QUALITY IS JOB 1.

Ribosomes are small as organelles go. They are in fact very large molecules (part protein and part a form of RNA) that attach themselves to a messenger RNA and follow its instructions, manufacturing precisely the protein molecule the gene is calling for. In most microscopes, you can't see any more of a ribosome than a dot or blob. There's a more detailed explanation of how ribosomes work in Chapter 2.

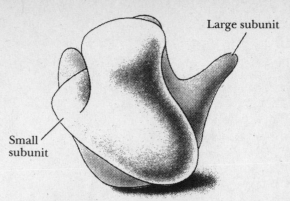

Large subunit

Small
subunit

*3.13 Like waltz partners, the two subunits of a ribosome have
arms that embrace each other and the strand of messenger RNA.*

THE ENDOPLASMIC RETICULUM—
PROTEIN PROCESSOR

It's one of the unwieldiest names in cell biology (scientists usually just call this component of a cell the E.R.), but
the full name may be appropriate since the thing itself is
also pretty unwieldy. It's a huge, floppy membranous bag
that's sort of folded up and wrapped loosely around the
nucleus, taking up maybe half the cell's volume. The E.R.'s
function is to provide a chamber, separate from the rest of
the cell's metabolic processes, for the chemical modification of many of the proteins made by ribosomes.

Cells need modified proteins because the proteins
made by the ribosome are not always in the form in
which they can do their job. It's kinda like adding up-
grades to your computer or your sound system to make
them do new things. If the simple, bottom-of-the-line
protein, as made by the ribosome, is intended for imme-
diate use, no E.R. is needed The ribosome makes the

Ribosomes   Membranes

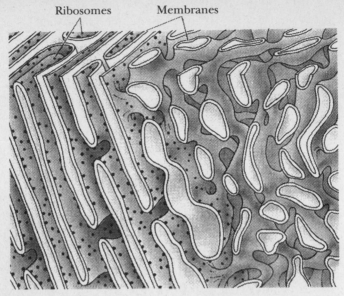

*3.14 The endoplasmic reticulum is an incredibly convoluted network
of membrane sheets that fills much of the space inside cells. Think of
it as a giant, deflated balloon that is repeatedly folded and wrinkled.
Black dots are ribosomes injecting their newly made proteins into the
ER's interior space where they undergo further processing.*

protein and simply turns it loose. Among these proteins
are those used as building blocks to make the cell's
"skeleton" and the tracks over which molecular motors
transport cargo within the cell.

But if the cell needs a deluxe model protein, the E.R.
is where the "chrome" and "pinstriping" gets put on.
Of course, you realize, cells are not really that frivolous.
What happens in the E.R. is sensible, necessary stuff:
sugars, phosphates, and other molecules that modify the
protein's overall shape and chemical properties are
attached in specific places. Many of these modified proteins are destined for jobs as enzymes either for use

within the cell or for export to other cells. For example, the cells of the pituitary gland, which is part of the brain, make the sex hormones that regulate the gonads. Those hormones are proteins that won't work without the requisite set of chemical attachments.

Proteins that need this customization have a special code at their beginning and end and once the ribosome cranks out the first segment of its protein (but has not finished making the whole protein), the newly made segment quickly binds to an access port into the endoplasmic reticulum. Like practically everything in life (at the cellular level anyway), processes like this one go forward because one molecule automatically fits perfectly into another ("hand and glove," "lock and key," pick your metaphor). The E.R. is studded with proteins that act as receptors for these special segments at the beginnings of proteins destined for modification.

When the receptor and protein glom onto one another, the protein segment is inserted into the endoplasmic reticulum. The ribosome, which stays outside the E.R., keeps on making more of the protein, and the newly forming molecule is literally injected into the E.R.

THE GOLGI APPARATUS—THE ONLY ITALIAN ORGANELLE

This structure, which looks like a stack of deflated balloons, was discovered by the Italian biologist Camillo Golgi (1844–1926) and is the only organelle named for a person. A typical cell may have a few dozen Golgi apparatuses, or simply "Golgi" for short. Their job is to make a few more necessary modifications of the proteins that were (first) synthesized by ribosomes and (then) modified by the endoplasmic reticulum. Golgi also sort and package proteins for shipment to other parts of the cell or out of the cell entirely. The exported proteins, for example, include digestive enzymes made in the liver but needed in the stomach, growth stimulants made in the brain to act on all parts of the body, and even antibodies, those infection-fighting proteins made by white blood cells. (There's a whole section on these guys in Chapter 4.)

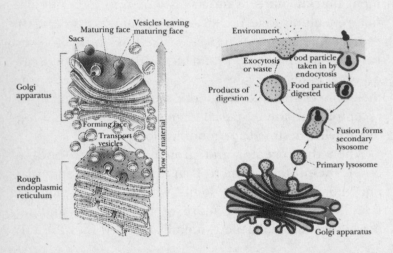

3.15 The Golgi apparatus, which you can think of as a stack of deflated balloons. After processing in the ER, proteins are ferried in vesicles to the Golgi for even more processing. Vesicles are moved by motor molecules (Fig. 3.9) and fuse as seen in Fig. 3.10.

Here's what happens. When the endoplasmic reticulum has finished its part of the job, a patch of its membrane bulges out and pinches off to form a vesicle. Inside it is a cargo of proteins that need further work. The vesicle then moves to the first of the "deflated balloons" and fuses with it, effectively dumping its cargo into the Golgi balloon. Enzymes inside the Golgi make changes to the sugars sticking out of the protein. Then another vesicle pinches off the first balloon and moves to the second balloon in the stack. Different changes ensue, then on to the third balloon, and maybe a fourth as well.

Just as important as the protein alterations are changes in the membrane of each vesicle—the addition of "address labels" that tell where the finished proteins are to be taken (hauled by the molecular motors we mentioned earlier).

Like all good manufacturers, cells put extra effort into the products intended for export. Many cells are in the export business, which in biology is called secretion. For example, pancreas cells make and secret insulin, dumping it into the bloodstream, which carries it to all other cells.

LYSOSOME—THE CELL'S STOMACH

We weren't kidding about cells being little organisms, little living beings in their own right. They eat, so naturally they have to digest their food. The body's stomach doesn't break down food to its smallest components. Lysosomes, which are filled with powerful digestive enzymes, finish the job and then send the broken-down food out to be used elsewhere in the cell. For example, the protein we eat in food is only partly broken down in the gut. Those protein fragments are carried in the bloodstream to cells elsewhere, which further reduce them to their constituent amino acids in lysosomes. The amino acids are then used by the cell as raw materials for the synthesis of new proteins.

The protein fragments enter the lysosome in vesicles that form at the cell's outermost membrane as a pit that bulges into the cell, essentially sucking in the protein fragments. As the pit on the membrane's surface deepens, it pinches off, breaks loose, and travels deeper into the cell until it fuses with a lysosome's membrane.

Lysosomes are sometimes called "suicide bags" because the enzymes they contain could digest the whole cell if they got out, which they do when the body is severely deprived of oxygen. During suffocation or drowning, cells become more acidic inside, which makes lysosome membranes break down and release their caustic juices. Brain cells are the first to undergo this reaction, often destroying themselves in four or five minutes after breathing stops.

CYTOSKELETON—
NO BONES ABOUT IT

Lacing their way throughout every cell are several kinds of networks of protein filaments, each made of many protein building blocks locked together. Some are straight lines, others more like lacework. Some cytoskeletons really are structurally vital, maintaining the cell's shape and helping it move. Others are the framework that support and move the hairlike projections, such as cilia and flagella, which appear on the outside of some cell types. Still other cytoskeletons act as the railways of the internal transportation system mentioned earlier— the one that molecular locomotives use to haul vesicles of cargo, such as the proteins processed through the endoplasmic reticulum and the Golgi apparatus.

The best-known cytoskeletal railway system is made of microtubules, which are tiny tubes (made of countless

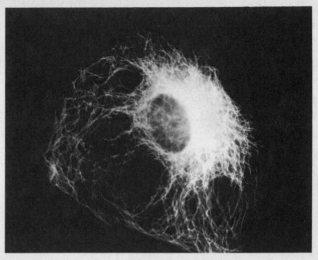

3.16 A cytoskeleton in every closet. This is a cell that has been stained with a chemical that binds only to keratin and glows so that its picture can be taken. Keratin, best known as the tough protein of skin and hair cells, is just one of several kinds of meshwork skeletons that exist in cells.

protein molecules that spontaneously self-assemble into that shape) that cluster around the nucleus and wind their way off to the distant cell membrane. At least two kinds of motor molecules haul cargo on microtubules— an uptown motor and a downtown motor. Really. One hauls cargo only from the nucleus outward; the other only operates in the opposite direction.

There is one type of cytoskeleton that everyone has seen. In fact, it is the only cellular structure that people pay money to see. It's made of a tough protein called keratin and it is the main type of cytoskeleton in skin. Deep inside, skin cells are alive; but as new skin cells divide, the older ones on top are pushed toward the outer surface of the

skin. The cells gradually make more and more keratin filaments, stuffing themselves and linking up with keratin in
neighboring cells. Then the cells die and virtually all other
components of the cells are lost, leaving only keratin. Hair
and nails are also the hardened keratin cytoskeletons of
dead cells. And, no matter what the shampoo ads say, you
can't feed dead cells.

THE MITOCHONDRION— THE ENERGIZER

Nearly all organelles do work that consumes energy. Sup-
plying that energy is the job of the mitochondria. These sausage-shaped structures take various forms of carbohydrates,
fats, and proteins (some of it prepared by the lysosomes),
extract the energy in their chemical bonds, and convert that
energy to a standard form the cell can use.

Food energy arrives in the cell in the form of molecules that are holding the energy in the bonds that keep

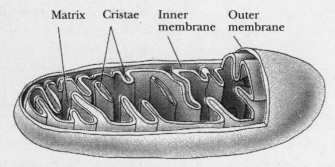

Matrix   Cristae   Inner   Outer
                       membrane   membrane

*3.17 You met mitochondria earlier as the stripey things in Fig.
3.8. These are the cell's energy processing and packaging cen-
ters. They take the energy from chemical bonds in digested food
and repackage it into the form of ATP, a molecule that can
carry energy anywhere in the cell and deliver it as needed.*

their constituent atoms together. The mitochondrion's enzymes break those bonds to release the energy but then immediately recapture it to form new bonds in a different molecule, called ATP, or adenosine triphosphate. ATP's advantage is that it can be distributed throughout the cell, and that all other organelles can easily break it down to extract the energy. ATP is like a universal battery that fits all kinds of devices.

Typical cells have scores of mitochondria but those with extra energy needs, such as muscle cells, are jammed with many thousands.

The reason we breathe in oxygen is because our mitochondria need it to "burn" the food molecules and extract the energy. This process is a form of combustion, much like a slow fire.

Mitochondria are unusual among organelles because they may be a form of helpful parasite. There is evidence that in the dim evolutionary past, mitochondria were bacteria that invaded cells and took up housekeeping. They are the size of bacteria and, like them, they have their own DNA (loose, not in a nucleus), their own ribosomes, and they reproduce within cells by dividing.

CELL MEMBRANE—THE OUTER LIMIT

Though often imagined as a mere bag holding all the organelles inside, the cell's outer membrane—also called the plasma membrane—plays a very active role in controlling the entry and exit of molecules. In an intestinal cell, for example, the membrane decides which products of digestion are absorbed into the bloodstream. Membranes of brain cells control the release of neurotransmitters, molecules that relay signals among neurons.

The main part of the membrane is made of trillions of small molecules whose chief attribute is that they have heads that attract both one another and water (hydrophilic) and tails that repel water (hydrophobic). Dump a bunch of these molecules into water, and they spontaneously assemble themselves into two-layered membranes enclosing droplets of water, like vesicles. The water-repellent tails immediately orient away from the water and toward one another, while the water-loving heads face out to the water inside and outside the membrane.

Water

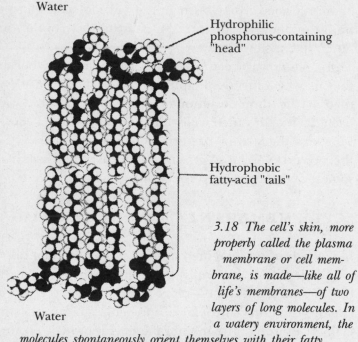

Hydrophilic
phosphorus-containing
"head"

Hydrophobic
fatty-acid "tails"

3.18 The cell's skin, more properly called the plasma membrane or cell membrane, is made—like all of life's membranes—of two layers of long molecules. In a watery environment, the molecules spontaneously orient themselves with their fatty tails (water repelling) together and their phosphate heads (water attracting) to the outside. Embedded in the membrane are many protein receptors.

Water

The hydrophobic inner layer is what holds the cell membrane's second major component—large protein molecules—in place. These proteins, which stick out both sides of the membrane, like doorknobs, control traffic into and out of the cell. They are receptors, or docking sites, into which will fit only the molecules that the cell wants to take in. When the right molecule comes along and docks with its receptor, the receptor protein pulls it into the cell. Receptor proteins stay in the membrane because their own hydrophobic and hydrophilic regions match up with those of the membrane and are naturally attracted to one another.

Along with receptors, the cell membrane is equipped with a variety of channels and ports, which pump specific atoms, such as calcium and sodium, in or out of the cell. These, too, are made of special proteins.

Despite the attractions that hold membranes together, they are rather fluid and the molecules slip around one another. When a newly made receptor protein enters the membrane, the membrane molecules make room as if they were a layer of floating Ping-Pong balls accommodating a block of wood. Cells have many kinds of receptors studding their membranes and sometimes thousands of each kind. The combination of receptor types in a cell membrane determines which molecules are allowed in and, since these molecules may trigger genes into action, they share in control of the cell's behavior.

MITOSIS: SEX AND THE SINGLE CELL

Actually, single cells (whether free-living organisms or members of a multicelled colony) don't have sex. They reproduce by the less pleasurable but still fascinating nonsexual process of cell division, which includes the process of mitosis. Actually, mitosis is among the most profound phenomena in nature. One unit of life transforms itself into two units of life. An aged cell becomes twin youthful cells. It is a process teenagers would envy: once it's done, there are no parents around.

Moreover, it happens millions of times every second in the human body, gradually replacing every cell except for those in muscle and nerve. The fastest to turn over are cells lining the intestinal tract, which live for only a day or so. (Who could last longer under those conditions?) Your entire skin renews itself in periods of weeks to months, depending on thickness. When skin is cut, however, the process speeds up to heal the wound. Then it returns to normal pace. Cancer may be caused if the regulatory mechanism that keeps cells dividing at a measured rate goes awry.

Cell division involves an elaborate series of closely coordinated steps in which one-of-a-kind structures, such as the nucleus and its chromosomes, are duplicated within the mother cell and each copy distributed to a daughter cell. Structures that already exist in multiples are simply apportioned to the daughter cells. The mitochondria, as we already said, undergo their own division, splitting as if they were miniature cells themselves, before being distributed to the two new cells.

The whole process starts with mitosis—the manufacture of a duplicate nucleus. The full set of 50,000 or so genes must be duplicated so that each of the two new cells can have a complete copy.

During DNA replication and cell division all of the cell's chromosomes change from the usual threadlike

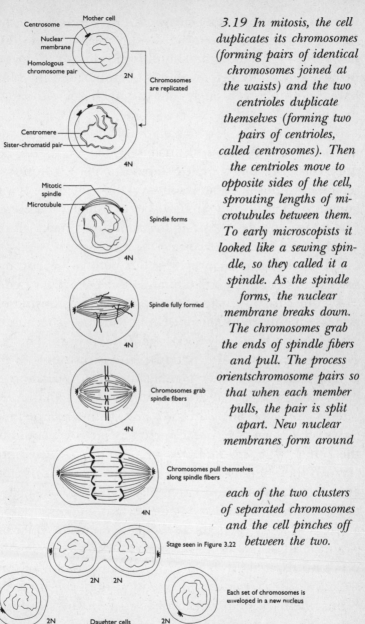

3.19 *In mitosis, the cell duplicates its chromosomes (forming pairs of identical chromosomes joined at the waists) and the two centrioles duplicate themselves (forming two pairs of centrioles, called centrosomes). Then the centrioles move to opposite sides of the cell, sprouting lengths of microtubules between them. To early microscopists it looked like a sewing spindle, so they called it a spindle. As the spindle forms, the nuclear membrane breaks down. The chromosomes grab the ends of spindle fibers and pull. The process orients chromosome pairs so that when each member pulls, the pair is split apart. New nuclear membranes form around*

each of the two clusters of separated chromosomes and the cell pinches off between the two.

form in which they can issue their commands. Now, they shrivel and coil into inert blobs—the "fat Xs" and "little sausages" of Chapter 2. The chromosomes stay in this inactive form until cell division is complete and the blobs have been delivered safely into the nuclei of two daughter cells.

Because of these form changes, all the genes are out of commission for hours at a time. This means the cell must carry out the complex steps of making a duplicate set of chromosomes and splitting the cell in two with genetic instructions that have been issued in advance. The genes do this by issuing instructions to make all the requisite enzymes in advance. Once a complete stockpile of enzymes has been prepared but before the chromosomes condense, the DNA strands must be replicated.

At some unknown signal, a group of DNA-copying enzymes—mainly one called DNA-polymerase (because it makes new DNA polymers)—go into action. The enzymes first unwind and split apart the two strands of the double helix. Then they manufacture a new partner for each, using the old strand as a guide, or template.

This works because the two strands are complementary—each is, in a sense, a mirror image, or a mold, of the other. (We covered this in Chapter 2, but if you skipped that, here goes again.) All single strands of DNA are made of four kinds of repeating units, or nucleotides, abbreviated A, T, G, and C. Their chemistry allows a single strand to contain any sequence but the opposite strand must always be of a complementary sequence. In other words, if there is an A at a given position on one strand, only a T can go into position opposite it on the complementary strand. And vice versa. In the same way, G and C are a complementary pair.

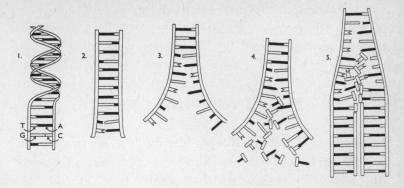

3.20 DNA replication turns one strand of genes into two identical strands of genes. In Step 1 we see the familiar double helix, but to make the process easier to follow, we've flattened it out to show the complementary rungs of A plus T and G plus C. Note that because of differences in shape, an A can link only with a T and a G only with a C. In Steps 2 and 3, the pairs unzip. In Step 4, loose subunits of DNA (called nucleotides) link up with their complementary partners. By Step 5, we see that two identical strands have been formed. Actually, the unzipping and synthesis of new strands is carried out by an enzyme called DNA polymerase.

DNA replication is carried out in a soup of loose nucleotides. As some enzymes are unzipping the double strands (taking down their genes, as it were), others are bringing loose nucleotides and fitting them into position to re-create the missing complement for each of the old strands. Because accuracy is essential, DNA-polymerase proofreads each new strand to see that it has the right nucleotides in each position. If not, it automatically snips out the errant nucleotide and replaces it.

A human cell, containing 50,000 or more genes made of an estimated three billion nucleotide pairs, takes about seven hours to make one copy of all its genes. Just ask your neighborhood copy shop to do three billion pages in that time. At this point, the cell has not the normal human set of 46 chromosomes but 92.

SORTING THE SOCKS

Now comes a new problem—sorting the chromosomes so that each new cell gets only one complete set. Shorting one of the cells would be fatal for it. To do the job right, the cell constructs a machine (made of interlocking proteins that spontaneously assemble into the right form) that surrounds the 92 chromosomes, reaches in from opposite sides, and appears to grab the 46 members of each set and pull them apart. The machine, called a spindle, will transport the sets of chromosomes toward opposite ends of the cell, and a new nuclear membrane will form around each. The process is like two blindfolded people reaching into a drawer with 92 mixed-up socks (from 46 pairs) and each person pulling out only one member of each pair.

How does the spindle know which 46 chromosomes to take? (This is the kind of intimate detail that cell biologists grapple with as they establish the fundamental mechanisms of life, and the answer emerged only in the 1980s.) Sock-sorting success is the result of a remarkable interaction between the condensed chromosomes and two mysterious organelles called centrioles. Nobody knows much about centrioles. Normally, they hover near the nucleus until time for mitosis. Then they move apart (nobody knows how) and take up positions on opposite sides of the cell. This happens as the chromosomes are condensing and the nuclear membrane breaks down.

Then each centriole sprouts long fibers in all directions. This is the spindle, called that because it looks vaguely like a weaver's spindle. The spindle fibers, reach into the mess of chromosomes, attach to a chromosome, and pull.

The simple secret of chromosome sorting is that the two members of each pair are stuck together at the middle—like socks paired and stapled somewhere near the heels. Only chromosome "staples" are clever little machines. They have the molecular equivalent of hands that reach out from each sock. When a spindle fiber comes near a staple, a hand grabs hold. This immediately orients the pair of socks—er, chromosomes—so that the opposite hand is reaching toward the opposite spindle and its fibers. When both hands have grabbed fibers, they start pulling—in opposite directions. The hands literally pull the pair apart and grapple along the spindle fiber, towing their chromosomes behind. In about fifteen

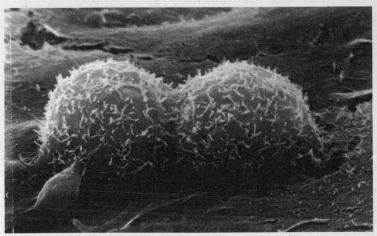

_ 3.21 *Caught in the act of dividing are these two cells. They were one, but it duplicated its chromosomes (mitosis) and began pinching off into two separate life forms.*

minutes, the chromosomes are separated and a new nuclear membrane forms around each set.

That's mitosis, but the cell still hasn't divided. As the two new nuclei form and their chromosomes begin unraveling to return to the threadlike state, a belt of long filaments, called the contractile ring, begins cinching down around the middle of the cell. The belt is essentially a tiny muscle made up of the same filaments that bulge biceps. Like a shrinking belt—positioned between the two new nuclei—the ring squeezes the mother cell's membrane until it pinches in two. Where there was one unit of life, there are—approximately one hour after the chromosomes condensed—two.

THE CELL DIVISION SISTERS: MEIOSIS AND MITOSIS

Why they needed such confusingly similar names, nobody knows. Except that there are similarities.

Mitosis (my-TOE-sis) is the duplication of a cell's complete set of chromosomes for regular cell division. One full set of chromosome pairs becomes two full sets of chromosome pairs. Each set goes to one of the two newly formed cells. This is the process by which single-celled organisms reproduce and by which cells in multicelled organisms multiply, so the critter can grow.

Meiosis (my-OH-sis) is different in that one cell duplicates its chromosomes (as in mitosis) but then divides twice to make four cells. If you followed the arithmetic, you'll understand that each of the four resulting cells has only a half-set of chromosomes (one member of each original pair). The reason is that the cells that do this are in the testes or ovaries and the results are going to be sperm or eggs.

SUMMARY

⏱ The cell is the fundamental unit of life. All living things are made up of one or more cells.

⏱ Within the human body—or any other body—most cells perform specialized roles, exercising only a small fraction of their genes (the ones needed to do a particular job, depending on which organ and tissue the cell is in).

⏱ Inside a cell, everything that happens is the result of chemical and physical reactions. Many of those reactions take place in specialized structures, called organelles, each of which is a small laboratory specializing in certain processes.

⏱ All cells in all species work essentially the same way. (There are slight differences in bacteria because they lack true organelles.)

⏱ The chief organelles are these:

Nucleus—Contains the genes, whose instructions to the cell are issued in the form of transcribed copies called messenger RNA. (This was a key point of Chapter 2.)

Ribosome—Reads the instructions carried by messenger RNA and manufactures the specified protein. (Covered also in Chapter 2.)

Endoplasmic reticulum—Receives some newly made proteins and modifies them by adding sugars and phosphates needed to make the protein fully functional.

Golgi apparatus—Modifies proteins further, sorts and packages them for shipment to other organelles within the cell or for export (secretion) from the cell.

Lysosome—Receives constituents of food from the bloodstream and breaks them down to their most basic constituents for reuse by the cell.

Cytoskeletons—Serve two basic roles. Some are structural supports, analogous to the human skeleton, allowing the cell to crawl around. Other cytoskeletons are the tracks of a transport system within the cell.

Mitochondrion—Takes carbohydrates, fats, and proteins (some digested by lysosomes), extracts the energy in their chemical bonds, and stores it in the bonds of ATP, the cell's universal energy source.

Cell membrane—Encloses the whole cell but contains various receptors and channels that control the movement of molecules into and out of the cell.

 Cell division (the way single-celled organisms reproduce and the way cells within bodies reproduce) requires, first, that the chromosomes be duplicated (the process of mitosis in which each strand of the double helix serves as a template for the replication of the opposite strand). Then the cell pinches in two between the duplicate sets of chromosomes.

CHAPTER 4

THE HUMAN BODY:
EXAMINING ENTRAILS
FOR THEIR OWN SAKE

YOU MUST REMEMBER THIS

The human body works fundamentally the same way as do the bodies of all other vertebrates (except maybe for Michael Jackson). The basic functions of life—such as sensing the environment, moving around, breathing, feeding, reproducing, and just hanging out—are all performed in more or less the same ways by the same kinds of organs in a salamander, a whale, and a human.

"Say! These little critters are not asexual after all."

IN THE BEGINNING

I t's not clothes that make the man . . . or the woman.

FERTILIZATION: WE HAVE TO START SOMEWHERE

Once during each woman's menstrual cycle, an egg leaves one of her ovaries (somehow, amazingly, the two ovaries take turns) and journeys down the fallopian tube toward the uterus, emitting chemical signals (let's just go ahead and call it perfume) to catch the attention of sperm cells. The human egg, though a thousand times bigger than most cells, never goes out alone. It is escorted by thousands of tiny cells that enclose it in a cloudlike "cumulus."

Nestled within its retinue, the egg basically surfs toward its rendezvous, riding waves created when cells lining the fallopian tube lash their hairlike cilia in coordination. The perfume molecules, being smaller, move ahead faster. If any sperm got lucky and entered the woman's reproductive tract, they sense the fragrance and swim, by the millions, toward the egg, lashing their long tails furiously. The first ones to arrive, however, are not so lucky. The cumulus is in the way.

The head of each sperm carries a chemical warhead that bursts open on contact, releasing enzymes that break down the "glue" holding the cumulus cells together. As more sperm batter the cumulus, its cells come loose and expose a second protective layer—a jellylike coating called the *zona pellucida*. No sperm is allowed through this coating unless it knows the password. It's a molecular password, of course, a protein of just the right cut that the

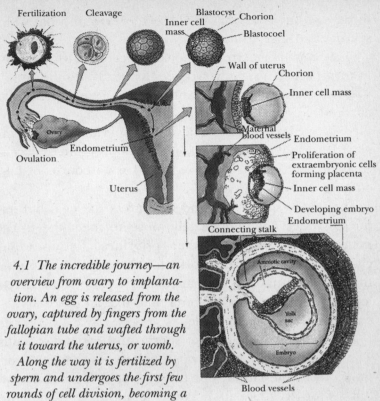

4.1 *The incredible journey—an overview from ovary to implantation. An egg is released from the ovary, captured by fingers from the fallopian tube and wafted through it toward the uterus, or womb. Along the way it is fertilized by sperm and undergoes the first few rounds of cell division, becoming a blastocyst, which invades the uterine lining, or endometrium. Once outer cells have made contact with the mother's blood supply, development of the true embryo begins and involves only two layers of cells deep inside. The layers are visible in the last diagram, between the amniotic cavity and the yolk sac.*

hopeful sperm sports on its outer surface. In the zona are the egg's equivalent proteins, shaped precisely to fit—lock and key, hand and glove (hey, this is sex, after all). One molecule fits into the other only if the two come from the same species. This choosiness is a vestige of the evolutionarily ancient and emotionally less exciting days when fertilization took place outside the body, in open water. Under those conditions an egg can't be too careful, what with sperm of all species drifting around.

Once egg and sperm have locked receptors, it's too late to turn back. The sperm wriggles its head into the zona. The two membranes touch and begin to fuse. Electrically charged atoms enter the egg, changing its charge and immediately blocking access by other sperm. The flip in electrical charge also triggers the egg's program of development, starting the fertilized egg on its course of becoming a new individual.

The whole sperm, tail and all, enters the egg, but its nucleus does not immediately move to meet the egg's nucleus. The egg has other priorities. Because the charge-caused blockade works for only a few seconds, the egg must make a permanent barrier. So the egg releases an enzyme into the zona that, like an epoxy hardener, makes it impenetrable.

Still the genetic union of sperm and egg must wait. The sperm carries the proper half-set of 23 chromosomes from the father, but the egg, at this point, has a full set of 46 (23 pairs) from the mother. While the sperm cools its heels, the egg's chromosome pairs split apart and one member of each pair is packed into a new, tiny nucleus and ejected from the egg. Biologists used to think that after all the foreplay, the two nuclei joined at this point, forming, at last, a one-celled embryo whose nucleus contained the requisite full set of chromosomes. But this never happens. Instead both egg and sperm, now each possessing just a half-set of chromosomes, go through a strictly choreographed ballet. First, each duplicates its 23 chromosomes, producing 23 pairs for each, or a total of 92 chromosomes. Then, as in ordinary cell division, the chromosomes cross combine—

ESOTERIC TERMS
(es-ə-'ter-ik tərms)

Egg: The reproductive cell made by females to join with the male's sperm during fertilization. Before fertilization, the human egg carries a complete set of 46 chromosomes. Upon fertilization, half will be discarded. A human egg is a thousand times bigger than a human sperm.

Chromosome: A long strand (made of the chemical DNA) of genes linked end to end.

Gene: A segment of a chromosome (made of DNA) that carries the instructions to tell a cell how to make one kind of protein molecule.

Sperm: The reproductive cell made by males to merge with the female's egg. It swims with a long, lashing tail and carries one member of each chromosome pair required by a body cell.

Pre-embryo: The entity formed by the union of sperm and egg, up until about Day 15. This is often called the "embryo," but that's confusing because for the first two weeks after sperm and egg unite, the main job of the resulting clump of cells is to make the placenta, not the baby.

Embryo: The group of cells destined to become the baby, from Day 15 up until the eighth week after fertilization.

Fetus: What the embryo is traditionally called after the eighth week. The significance of eight weeks is that this is the time by which all the major organs and anatomical structures have been started, at least in rudimentary form.

each member of Dad's 46 joining up with the corresponding member of Mom's 46. Then, the two sets of 23
pairs move apart and, finally, become two nuclei (a full
set of 46 chromosomes for each). At last, the cell divides,
producing the two-celled stage.

MAKING THE PLACENTA IS JOB ONE

The process of human development up to this point
takes about a day and will take another two weeks before
the appearance of the first structures destined to become
the baby. Thus, there is no one moment that can be
called "conception." Instead, the launching of a new
generation is a process that unfolds in phases. Until the
first embryonic structures appear, the conceptus or pre-
embryo, as developmental biologists call it, must keep its
priorities straight. The pre-embryo has to construct the
placenta. Only after the pre-embryo has burrowed into
the mother's womb wall and tapped into her bloodstream will there be time and energy to start work on
the embryo proper. That, however, won't happen until
Day 15.

Once the pre-embryo's two founding cells begin dividing, however, the cells do start communicating with the
mother. They manufacture a hormone called HCG
(human chorionic gonadotropin) that signals the uterus
to get ready. Remember, the pre-embryo is still moving
down the fallopian tube; it will not be ready to invade
the uterine wall for another four or five days.

The two cells divide into four, each half the size of
their predecessors. Four into eight, halving in size again.
Eight into sixteen, smaller still—all still packed inside
the hardened zona pellucida. On Day 5 the pre-embryo,

now usually called a blastocyst, hatches from its shell, the zona. "Hatching" is actually the scientific term.

On Day 6 the blastocyst begins tunneling into the uterine wall. To do this, it secretes enzymes that break down the mother's tissues, invading in a process almost identical to that of a malignant tumor. (Seriously. Cancer researchers study embryo implantation for clues to cancer.) The blastocyst sprouts tentacles that penetrate deeper into the mother's tissues and wrap around the mother's blood vessels. This is so the blastocyst can extract oxygen and nutrients from the mother. On Day 9, after having established what is essentially a parasitic life-support system, the uterine wall heals over the blastocyst.

There are now hundreds of cells in the pre-embryo, and they begin to organize themselves into two hollow balls within the larger ball. The place where the two inner balls meet, like two balloons pressed together, is a double layer of cells called the embryonic disc. It is within this sheet of cells that the true embryo will begin to form, around Day 15—about the time the mother first misses a menstrual period.

JOB TWO: MAKING A BABY

Construction of the true embryo begins as a tiny pit opens in one of the two contiguous layers of cells. The cells bordering the pit crawl under their neighbors, creating and occupying a space between the two layers. The result is three layers. (See Fig. 4.2.) Over a day or so, the pit lengthens, forming a groove. Along the groove, moving sheets of cells from the top layer migrate into a middle zone. From these three layers will arise all the

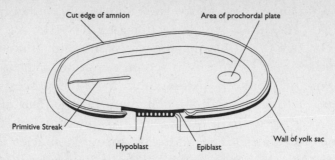

*4.2 Humble origins: The embryo that became you started as
two layers of cells. This diagram shows just the two layers, having
cut away all the other stuff seen in the last step of Fig. 4.1. The
first step in development of the true embryo, the primitive
streak, is not something college freshmen do but is the change
in cell position that lays down the rudimentary spinal cord.
Cells at the edge of the streak move between the two layers,
forming a middle layer and a groove (which covers over to
become a tube) that defines the embryo's axis. The backbone
will form around this tube.*

organ systems as various patches of cells bend, fold, and
slither into new positions.

It has taken two weeks from fertilization to reach this
primitive stage when the embryo is nothing more than
three layers of cells. Yet within the next week, the embryo will form a head and a tail, the rudiments of a
brain and a spinal cord, the beginnings of eyes and ears.

This is a time of rapidly growing complexity. At various
points, certain groups of cells get up and crawl off to new
homes. On arrival, the cells settle down and, like a young
couple moving to suburbia, they proliferate. Sometimes
whole sheets of cells slither away in unison. Each new layer
or fold creates a new environmental context that triggers new
combinations of genes into action. As a result, cells become
specialized for different roles in the body.

4.3 At 20 days of development, the true embryo continues to form as sheets and ribbons of cells roll up into special shapes. Here we're looking down on that embryo that we saw from an angle in the previous diagram.

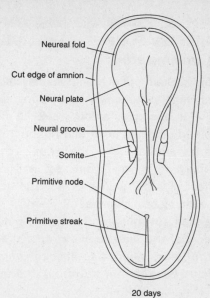

Neureal fold

Cut edge of amnion

Neural plate

Neural groove

Somite

Primitive node

Primitive streak

20 days

Certain ribbons of cells, for example, curl themselves into a network of tubes to make the circulatory system, and at one point along one tube some of the cells change again as muscle genes are activated. The heretofore undifferentiated cells become specialized; muscle cells arise and start contracting rhythmically. Working in concert, they become a primitive, one-chambered heart, pushing the embryo's own blood into the placenta and back. At this stage—three weeks after fertilization—the embryo is the size of a sesame seed.

HEREBY HANGS A TAIL. GILLS, TOO

It's true. At certain stages, human embryos have tails and gill pouches (structures that in fish become true gills). In humans, however, both shrink away long before

HEREBY HANGS A TAIL. GILLS, TOO
(*continued*) . . . .

birth. Such curious anatomical embellishments bespeak ancient evolutionary ties among all animals.

For example, a four-week-old human embryo has a clearly defined head and tail and buds that will become arms and legs. Yet the human embryo is indistinguishable from an embryonic pig or lizard or bird or any of a host of other animals. They too have gill pouches.

Biologists assume all animals have similar sets of genes governing early stages, but as development proceeds, the inappropriate structures disappear because the genes needed to change them into mature organs have been lost or have mutated into different genes.

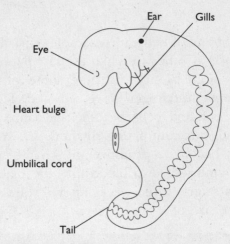

4.4 Time passes. Barely over a week in this case and the embryo that was a mere two layers of cells is now, 28 days after fertilization, a fairly complex organism with the rudiments of eyes, ears, a heart and, of course, a tail and things resembling gills. The whole thing is the size of a lentil.

WHO'S IN CHARGE HERE?

How do the embryo's cells know what to become, especially since all cells carry the same set of genes?

The answer, if you recall from Chapter 2, is that genes are lazy. They do nothing until they hear from environmental factors. Specific molecules from outside need to reach the cell to trigger specific genes into action. At every position within an embryo, each cell is receiving molecules from various sources that trigger (or suppress) the work of its genes. The combination of incoming signals determines which genes are to respond. In other words, the signals from outside the cell tell the cell what to become.

Studies of animal embryos, for example, show that cells at the head end of the three embryonic layers of cells secrete certain proteins that diffuse toward the tail end. Other chemicals diffuse from tail to head or dorsal side to ventral. Since the concentrations vary from place to place within the embryo, diminishing with distance from the source, all a cell has to do is sense how much of each signaling chemical is coming in and it will have a three-dimensional coordinate system telling the cell exactly where it is. Of course, cells don't actually think about this. It's just that a combination of various concentrations activates a certain set of genes. In one part of the embryo, the signaling cocktail triggers the genes needed to grow an arm. In another place, a different mix says, "Be a gallbladder."

Early embryos are equal-opportunity environments. All cells have equal potential to become anything—a brain cell or an anal sphincter cell. But as specializations develop, the cell builds a résumé that absolutely limits fu-

WHEN DOES LIFE BEGIN?

Few questions have generated more agony in societal debate. And yet the way this one is posed misses two key points.

First, life began only once, billions of years ago, and has been relayed via reproduction from one generation to the next ever since. The egg is human (not, of course, "a human") and alive and if it does not get fertilized, it must die. Each sperm is human and alive and if it does not fertilize an egg, it, too, must die. The few sperm and eggs that unite relay life to a new generation.

Second, the question should really be phrased: When does personhood begin? Or when, in the developmental process, does the entity (pre-embryo, embryo, fetus, or baby) qualify as a person—under the law or under your personal moral code?

ture job opportunities. As cells express various specific sets of genes, others are permanently switched off. One by one, career paths are closed—except for one. And in that job, cells have lifetime appointments.

During the first eight weeks of gestation, the cells organize themselves into rudimentary versions of all the major organ systems. This completes the embryo phase. From this point on, doctors refer to the developing human as a fetus. Now development consists of adding details to each of the organs (specialized tissues within them), developing interconnections among tissues and organs, changing the proportions of the body, and just simply growing larger.

It's not nearly as dramatic as the embryonic period

but it does make a big difference. For one thing, without fetal development for the next seven months, we'd stand only a little over an inch tall. The result of nine months of construction is perhaps the most astonishingly complex assemblage of interconnected subsystems conceivable—a baby. And not just a baby. It is a living body that will continue to develop for years and, in most cases, continue functioning with relatively little maintenance for decades, sometimes even a century or more.

BIG EVENTS IN LITTLE LIVES

Day 1 Fertilization.

Day 6 Pre-embryo begins burrowing into womb wall.

Day 11 Pre-embryo makes contact with mother's bloodstream.

Day 15 (1 mm long) First structure of true embryo appears.

Day 22 (2.5 mm) One-chambered heart starts beating.

Day 27 (3.5 mm) Arm and leg buds appear.

Day 30 (5 mm) Primitive eye visible.

8th week (20 mm) All major organ systems started. Fingers are separate. Arms and legs move but are not linked to brain. Brain cells are not connected to one another.

9th week (40 mm) Male and female genitals begin to look different.

10th week (50 mm) Face takes on human cast.

17th to 20th week "The quickening," when mother can first feel fetus moving.

20th week Brain cells start forming synapses—connections that allow signals to be transmitted. But cortex, the conscious part, is not wired to the rest of the body.

5th month First links between brain's cortex and rest of

BIG EVENTS IN LITTLE LIVES (*continued*)
. . . .

the body. Until this happens, the thinking, feeling part of the brain gets no signals from the rest of the body.

7th month Brain activity, as measured by brain-wave monitors, begins to resemble that of a fully developed baby.

9th month Birth. Brain development will not be complete for many months. Reproductive system still immature. Attitude toward parents remains ungrateful until at least middle age.

THE MEAT MACHINE

THE CIRCULATORY SYSTEM: YOU GOTTA HAVE HEART AND BLOOD

The bottom-of-the-line, one-chambered heart that the embryo made in Week 3 (similar to what insects and mollusks have) is not nearly good enough to meet the needs of a large mammal. During gestation the primitive heart adds three more chambers. It goes through a two-chambered phase (like a fish's heart), then a three-chambered stage (like the hearts of amphibians and most reptiles) before finally achieving the four-chambered heart that is standard equipment for birds and mammals.

The heart's job, of course, is to pump blood through the lungs (where the red blood cells pick up oxygen) to the rest of the body (where they trade oxygen to cells for carbon dioxide, a waste product of cellular metabolism). The CO_2 is taken back to the lungs, where it diffuses into

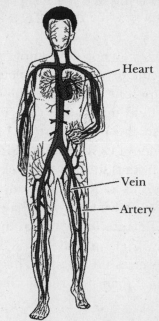

Heart

Vein

Artery

Cardiovascular system

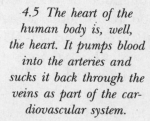

4.5 The heart of the human body is, well, the heart. It pumps blood into the arteries and sucks it back through the veins as part of the cardiovascular system.

the air that will be exhaled. At the same time, the blood also passes near the intestines to absorb nutrients from food—raw materials that are delivered to cells throughout the body for use in making new proteins and other substances.

Arteries carry blood away from the heart; veins, back to it.

The blood also contains a goodly part of the immune system—the so-called white blood cells, which have many jobs, including gobbling up invading bacteria and dead and dying cells of the body (including the 100 billion red blood cells that expire daily).

THE LYMPHATIC SYSTEM: THE MOST OBSCURE ORGAN

You may not even know you've got one of these. It's a circulatory system completely distinct from—but linked to—the one that carries blood. It carries lymph, the watery fluid that bathes cells throughout the body. Lymph is also the liquid part of blood (constituting about 60 percent of its volume) and in that role, it's called plasma. When this stuff turns yucky, like when it oozes from sores, it's called pus. Lymph contains white cells, also known as lymphocytes, that are part of the immune system. In infected wounds, the lymphocytes gobble up so many bacteria that the cells grow fat and literally eat themselves to death. This is what makes pus more opaque than the usually clear lymph.

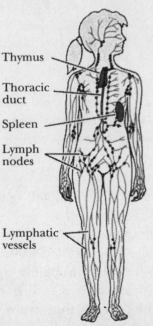

4.6 The "other" circulatory system is the lymphatic system, which consists of tubes that collect the liquid part of blood that has seeped out of the cardiovascular system, filters it through the lymph nodes and dumps it back into the blood stream.

Thymus

Thoracic duct

Spleen

Lymph nodes

Lymphatic vessels

About a gallon of lymph (plasma) seeps out of an adult's blood vessels every day. Often this is for good reason—to bring lymph-borne immune cells nearer damaged or infected tissues. Extra seepage is common in inflammation and causes swelling. Tissues would keep puffing up, in fact, if it were not for the open ends of the lymphatic system's smallest vessels. The lymph flows into these and is collected into one major pipe that pours it back into the bloodstream near the neck. On the way, however, lymph passes through lymph nodes, which are also key parts of the immune system. This is where many immune system cells reside, including the ones that make antibodies and release them into the lymphatic system, eventually to enter the bloodstream.

Unlike the cardiovascular system, which has a heart, the lymphatic system has no central pump. Instead, it depends on the flexing of whatever muscles happen to be nearby to squeeze its vessels. Lymphatic vessels have one-way check valves, so each flex of an adjacent muscle pushes the fluid in one direction only. This explains why your feet swell up when you've been standing all day. There's not enough muscle action to squeeze the lymphatic vessels and the lymph pools in your feet and legs. Walk for a short length of time and the action keeps the lymph pumped out.

THE RESPIRATORY SYSTEM: IT SUCKS

You'd think the breathing apparatus would be pretty simple. Two lungs connected to pipes that carry air in from the nose and mouth and then let it out. But, in fact, experts debate precisely how we breathe. The air flows just as you might think, but the hard part is why

the air moves. Two main sets of muscles are involved—
the diaphragm (which forms the bottom of the chest
cavity, separating the chest from the abdomen below)
and the muscles between the ribs (the stuff that's so
good barbecued). The main force, everybody agrees,
comes from the diaphragm. You breathe in because the
diaphragm pulls down, receding into the abdominal cavity and enlarging the chest cavity. At the same time, the
rib muscles contract, lifting the ribs, which are hinged
at the places they join to the spine. As a result, the chest
expands, sucking air into the lungs, like pulling out the

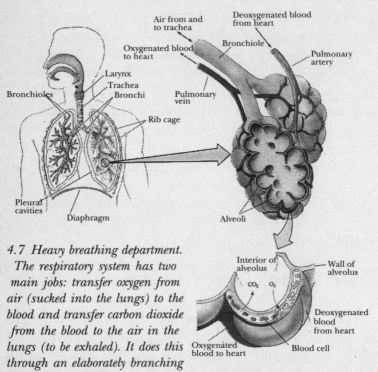

*4.7 Heavy breathing department.
The respiratory system has two
main jobs: transfer oxygen from
air (sucked into the lungs) to the
blood and transfer carbon dioxide
from the blood to the air in the
lungs (to be exhaled). It does this
through an elaborately branching
system of passageways that bring air into tiny alveoli, where it is
just one cell layer away from the blood stream.*

piston on a syringe. The debate is over exactly which rib muscles do what. Some are involved in breathing out and anatomists can't be sure which do what.

Breathing out, on the other hand, is mostly a passive action. The muscles that worked to suck in air simply relax, and the elastic chest tissues simply spring back, squeezing air out. Of course, the rib muscles and others (such as the abdominals) can be called into action to blow hard.

During one minute of breathing while not working hard, about a gallon of air enters and leaves the lungs. During heavy exercise, that airflow can increase to thirty or forty times as much.

The tubes that carry air into the lung divide many times, like the branches of a tree. The smallest branches are where the air comes closest to the bloodstream. In fact, the layer of tissue separating air from blood may be the thickness of just one cell, stretched flat. Those innermost cells of the lung relay oxygen and carbon dioxide back and forth. Unfortunately, they also relay garlic forth. The surface area that gases can cross in one person's lungs is about 800 square feet, or roughly the size of a badminton court.

MUSCLES: PUMPING PROTEIN

The muscles that we use to move around—called skeletal muscles because they run from one bone to another—are not the only muscles in the body. There are the muscles that contract just once at birth (closing off the umbilical cord and rerouting blood flow to the lungs), and there is the heart muscle, built to keep pumping for a lifetime. There are even muscles in your blood vessels that help keep the blood flowing. Humans can talk because of muscles that work the voice box and tongue. There are muscles in your kidneys that squirt fluids from one chamber to another. The stomach has muscles to mash the food and mix it with digestive juices. The intestines have muscles to keep the digested matter moving along. The uterus has muscles to push the baby out at birth.

The eyes are particularly well-endowed. There are muscles that blink your eyelids, and muscles that aim the eyes, and muscles that pull on the lens, stretching it to change its focus. The muscle that closes the iris (the part whose color means so much to lovers) is a belt that encircles the pupil (the hole in the iris where the light goes in). When the muscle contracts, the pupil shrinks. To open the pupil, there are other muscles positioned like spokes from the pupil out to the edge. They pull, and the pupil widens.

There is a tiny muscle attached to each hair that makes it stick up when you're cold—a vestigial effort to fluff up the fur for better insulation. There are even the same tiny muscles in follicles with no hair; they make goose bumps anyway.

There are thousands of muscles in each human body, and all flex the same way (though maybe with varying effect on the opposite sex). The work is done by two kinds of protein filaments inside each muscle cell: actin and myosin. These are long parallel filaments that lie next to each other in large numbers. Each myosin filament has lots of hands that reach out in all directions. Well, not hands, really, but molecules that can grab and pull. When a muscle contracts, the myosin hands grab the actin filaments and pull. The action is like that of a man in a canoe who grabs a dock and pulls his canoe forward. The hands let go, reach ahead, grab, and pull again. As long as the hands keep pulling (and reaching ahead at different times), the muscle contracts. Once

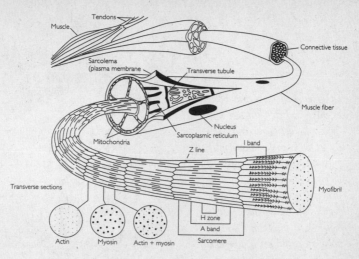

4.8 The force that is always with you basically comes from two kinds of long filamentous protein molecules called actin and myosin. Seen in the foreground as overlapping lines. Look at the last segment, which has been cut. See those little branches that look like fringe from some of the lines? Think of them as little hands reaching out from myosin filaments to grab the filaments without fringe, which are actin. When that happens the myosin hands pull, working hand over hand, and slide past the actin filaments. This makes the muscle get shorter and thicker, pulling the two ends together. When kazillions of actins and myosins do this, you're flexing your muscle.

they let go, the muscle relaxes. Muscle cells are a bit more complex than that, but that's the general idea.

At birth or within a few months after, Arnold Schwartzenegger had all the muscle cells he was ever going to have. That's because once his muscle cells became fully specialized during fetal development, they got so full of the machinery for generating a force (mainly actin and myosin) that they could no longer divide. Exercise simply makes the existing cells bigger. This applies to Danny DeVito, too.

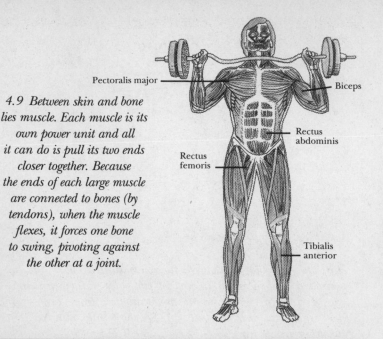

4.9 Between skin and bone lies muscle. Each muscle is its own power unit and all it can do is pull its two ends closer together. Because the ends of each large muscle are connected to bones (by tendons), when the muscle flexes, it forces one bone to swing, pivoting against the other at a joint.

Pectoralis major

Biceps

Rectus abdominis

Rectus femoris

Tibialis anterior

DARK MEAT OR LIGHT?

There are two ways to exercise and they build muscles in different ways. Low-intensity, long-duration exercise, such as long-distance running and swimming, builds up endurance without much gain in power. On the other hand, high-intensity, short-duration exercise, such as weight lifting, builds power but doesn't add to endurance. Both methods have their effects at the cellular level because there are two basic types of skeletal muscle fibers (aggregations of many muscle cells) and they get and use their energy in different ways.

Low-intensity, long-duration exercise causes muscle cells to produce more mitochondria, which make ATP (the energizer from Chapter 3) from food and oxygen. And this form of exercise encourages more blood capil-

DARK MEAT OR LIGHT? (*continued*)

laries to grow into the muscle, to bring in oxygen and food and to carry away waste products more effectively. These endurance cells are also rich in a protein called myoglobin, which is a little like the blood's hemoglobin in that it can take up oxygen, hold the oxygen, and release it on demand. Myoglobin is a short-term storehouse for oxygen within the cell. Like hemoglobin, myoglobin is red, and so muscle with a preponderance of these cells is dark in color. The dark meat of a chicken is rich in long-endurance muscle cells.

High-intensity, short-duration exercise works a different set of muscle cells, the white muscle fibers. They are, as astute readers will have guessed, predominant in the white meat. In humans, most muscles have a mixture of both types. But the back and legs, because they have to maintain the upright posture for long periods, do have more of the long-enduring oxidative fibers. Arm muscles, on the other hand, are richer in the ''white meat'' fibers that respond to weight training. With exercise, these muscle cells make more actin and myosin and pack them into the muscle fibers. This makes the muscle thicker. These ''power cells'' also use energy in a different way. They don't need extra oxygen since their work is done quickly. But they do need to tap lots of energy quickly. Mitochondria can't make ATP fast enough, so these cells have a second, faster way to produce ATP. This uses a carbohydrate molecule called glycogen, the form in which cells store the sugar glucose. Power cells contain lots of glycogen and, when they need a burst of ATP, they can, within seconds, break down the glycogen, and use its energy to make more ATP.

SKELETAL SYSTEM: PICKING BONES

Bones are alive. The body's framework, the levers on which muscles pull to let us turn the pages of books and perform other activities, are a living, constantly changing tissue.

Although bone is made mostly of hard mineral (calcium phosphate), it is shot through with soft, living cells that created the hard stuff. Slice through a bone (somebody else's, preferably), look under a microscope, and you will find individual cells sitting in tiny holes. They

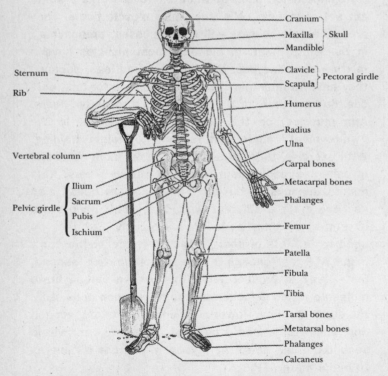

4.10 Out of the closet: The skeletal system.

are trapped in their spaces but not isolated. Each bone cell has several long tentacles that reach out through narrow passageways to make contact with tentacles from other bone cells.

There are two main kinds of bone cells—ones that destroy bone and ones that create it. This Siva-like nature is essential if bones are to grow. Think about this. Babies have little bones. Adults have big bones. How can a hard, rigid thing like a bone grow?

It is because some cells are chewing their way through existing bone, dissolving the mineral, and other cells are coming along behind and—like slugs leaving a slime trail—laying down a secretion that causes the dissolved mineral to crystallize again. The process is rather like what you see when they resurface a street. A machine comes along and chews up old pavement as it creeps along. And then another machine follows behind, laying down new pavement. To make a bone grow, there is more laying down of new bone than taking up of old.

At the end of puberty, hormones tell the bone-making cells to stop. But the cells don't go away. They just sit there, trapped inside their own creation and reaching out their tentacles, some of which make contact with blood vessels that also snake through the bone. If the bone breaks, these cells are called out of retirement and they start up again—tunneling their way across the break and laying down new bone to connect the two parts.

DIGESTIVE SYSTEM: ALL GUTS AND NOT GORY

Now don't get grossed out by this, but the first digestive organ to act on food is the mouth. Saliva is a digestive juice. It breaks down starch. Starch is made of sugars bound together. This is why the longer you chew a piece of bread, the sweeter it gets.

When your tongue forces a mouthful of food toward the back of your mouth, the nerves there respond by setting in motion a series of muscular activities that move your vocal cords in position to close the windpipe, lift the soft palate (to shut off access to the nasal passages), and activate a series of muscles that encircle your esophagus (food pipe) just above the food. Food doesn't fall into your stomach; it's pushed. As each ring of muscles squeezes in turn, the food is shoved along. This is why you can swallow while standing on your head, although Miss Manners discourages the practice.

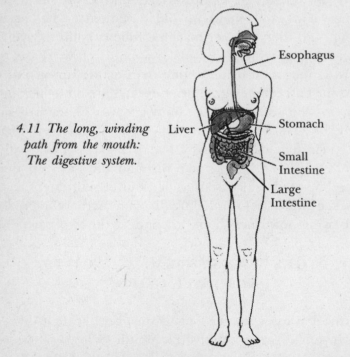

4.11 The long, winding path from the mouth: The digestive system.

Esophagus

Liver

Stomach

Small Intestine

Large Intestine

VITAMIN FUNK
. . . .

In the ninteenth century, it became fashionable in Asia to grind away the brown hulls of rice, leaving the grains a more appealing white. Soon after, a new disease called beriberi (Singhalese for "extreme weakness") became widespread. Symptoms varied, but the disease usually involve debilitating heart disorders.

Various researchers noticed that beriberi was common among eaters of polished rice but absent among the poorer social strata, where people made do with old-fashioned, less-desirable brown (unpolished) rice. In 1912 the Polish scientist Casimir Funk found that pigeons with beriberi could be cured by feeding them discarded rice hulls. Funk suggested beriberi and some other diseases were the result of certain nutrients missing from the diet. He dubbed them vitamins, for "vital amines."

Funk was wrong about their chemical nature (amines are substances derived from ammonia) but right about the existence of diseases caused by the absence of certain trace nutrients. The necessary factor in rice hulls was isolated in 1926—thiamine, now usually called vitamin B1.

The stomach exists because we can eat faster than we can digest and absorb the nutrients. The stomach stores the food and mixes it with a second digestive enzyme (for breaking down proteins) plus hydrochloric acid, which just plain dissolves a lot of stuff we eat. These are secreted by little glands in the stomach wall. Nutrients cannot be extracted until food is almost totally liquefied. Stomach muscles mash and mix it all. (If you feel a little

SCURVY
. . . .
During the long voyages of the Age of Exploration, sailors subsisting on salt pork and beans often came down with scurvy. It's a miserable disease, befitting its name. Symptoms include hemorrhaging and decay of skin and flesh, bleeding gums, loss of teeth, general debility and, eventually, death.

Around 1600, a British physician discovered that sailors who ate citrus fruits didn't get scurvy. He prescribed daily lemon juice for all British sailors. In 1865, the Royal Navy switched to limes, and its men have been known ever since as "limeys." In 1932, scientists found the active factor in citrus—ascorbic acid—and dubbed it vitamin C.

queasy and want to skip to the next section, that'll be okay.) Very little is absorbed into the body through the stomach wall. One exception is alcohol, which is why booze can hit you pretty fast.

The human stomach empties gradually (taking about four hours for a full load), squirting little dollops (really, it's okay if you want to skip a page or two) of stuff through a valve called the pylorus into the so-called small intestine. More digestive enzymes will be added here. Many are made by the pancreas, and one, called bile (formerly called gall and stored in what is still called the gallbladder), which breaks down fats, is made by the liver.

So, if these organs are making such potent digestive enzymes, how come they don't digest themselves away? Good question. What is made is not the working form of the enzyme itself but a precursor that is safe to handle. Only when the precursor gets to the gut does it meet another enzyme that converts it into the potent stuff.

The small intestine is small only in diameter when compared with the large intestine, which comes later. In an adult human, the small intestine is about twenty feet long and has so many convolutions that its total surface area is nearly 6,000 square feet, which is about one-eighth of an acre—the size of many city lots. When the food has been reduced to its fundamental components, these are absorbed into the cells lining the small intestine and relayed to blood and lymph vessels coursing nearby.

Waves of muscle contractions push the small intestine's contents into the large intestine, or colon. Two key events

VITAMIN X-CESS

So many Americans, especially food processors and advertisers, make such a big deal out of vitamins, you'd think they were the most important element in a good diet. Americans swallow so many vitamin pills and so much food "fortified" in vitamins that the world's richest potential source of these chemicals is now human urine and feces. The body simply dumps what it cannot use or store.

In fact, vitamins are needed only in trace amounts and are referred to as micronutrients. Far more troublesome in the American diet are the levels of macronutrients—proteins, fats, and carbohydrates, all of which must be eaten in quantities millions of times greater than vitamins. The leading diet-caused diseases in the industrialized world (heart disease, stroke, a form of diabetes, and possibly some forms of cancer) have nothing to do with vitamins, or even with shortages, and much to do with overconsumption of fats.

happen here: extraction of water from the highly liquid stuff it receives (turning the goop into, ahem, a more manageable form, unless, of course, you have diarrhea, which is usually the result of stuff moving through the colon too fast to get the water out), and the farming of a huge bacterial colony. The colon is home to a species of bacteria called *Escherichia coli*, or *E. coli*, for short. This microbe, which is the one most commonly used by genetic engineers when they want to tinker with genes, feeds on what it gets from the small intestine. But these bugs are not parasites; they're welcome inhabitants. They manufacture two essential nutrients: vitamin K and biotin, which are absorbed through the colon wall.

NERVOUS SYSTEM:
TOUCHY AND FEELY

In physical terms, the human brain is not so impressive. It weighs a mere 3½ pounds, most of it is water, and it is about the consistency of pudding. Yet, except for those of Saturday morning cartoonists, human brains are by far the most complex structures in the known universe. Even though the brain remains the most poorly understood organ in the body, the little that is known fills volumes and puts the subject far beyond the scope of this book. But we won't let that stop us from offering a hint or two.

The nervous system is divided into two parts, largely on a geographic basis. The central nervous system consists of the brain plus the spinal cord, which runs through holes in the backbones to the base of the spine. The spinal cord is really part of the brain and it actually does some decision-making without ever consulting the conscious part of the central nervous system up top. The peripheral nervous system consists of the nerves that run between the

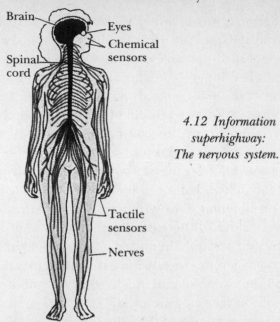

Brain

Eyes

Chemical sensors

Spinal cord

Tactile sensors

Nerves

4.12 Information superhighway: The nervous system.

spinal cord and everything else in the body. There are two types of these: sensory nerves and motor nerves. The sensory nerves collect information from the sensory organs— the eyes, ears, nose, tongue, and the skin. Each of these organs houses the tips of sensory nerves, which carry the information to the central system. The motor nerves carry signals from the central nervous system to the muscles, telling them to contract or relax.

Like all organs, the nervous system is made of cells. These are called neurons. The basic shape of a neuron consists of a central "body," holding the nucleus and other major organelles, plus a lot of long tentacles that branch repeatedly, producing hundreds of tips. Each tip makes contact with another neuron (at a place called a synapse) and can send signals to it. In turn, each neuron receives signals from hundreds or thousands of other neurons at synapses all over its surface. Typical neurons have several short tentacles and one very long one, called the axon. The longest human axons are those that run from the base of the spinal cord to the tip of the big toe. A nerve is simply a bundle of axons, like a telephone cable carrying wires from many phones.

Neurons transmit signals in two ways. Along the length of any given neuron, the signal is electrical—a pulse of current much like that in a wire. When the pulse reaches the tip of a tentacle (at a synapse where the tentacle comes very close to the surface of another neuron), the pulse causes certain chemicals to be released. These chemicals, called neurotransmitters, drift across the short space of the synapse and dock with receptors on the surface of the receiving neuron. That, in turn, triggers a new electrical impulse to flow through the next neuron.

Among the many jobs our brains do without even having to think consciously are to run the breathing muscles, send extra blood to the gut when there is food to absorb, tell the pancreas to deliver more digestive juices, stimulate the sweat glands to sweat, control the heart rate and, when the going gets tough, switch on all the reactions needed for the "fight-or-flight" response.

One of the simplest things that nervous systems do is handle reflexes. These are the snap decisions made in the spinal cord without consulting any higher-ups. For example, if you touch a hot stove, the heat causes a pain signal to travel along a sensory nerve in your arm to the spinal cord. The signal will be sent up to the brain for the consciousness to ponder, but because that takes valuable time, the spinal cord goes ahead and issues instructions to the motor nerves to contract certain arm muscles and jerk the hand away. The hand moves to safety before the conscious part of the brain ever "feels" pain and realizes what is going on.

Neuroscientists can understand these simple reflex systems fairly well. They may involve as few as two or three neurons. The human brain, on the other hand, is made of about 100 billion neurons, each with an average of about 1,000 synapses linking it to that many other neurons. That degree of interconnection is what makes the brain capable of all the extraordinary things it does. This is also what makes it nearly impossible to understand the brain, at least with today's level of scientific understanding. Somehow, out of all that rich circuitry, the brain generates love, music, courage, literature, and mud wrestling.

IMMUNE SYSTEM: THE DEFENSE DEPARTMENT

In a natural world teeming with millions of forms of life that have evolved to exploit every conceivable ecological niche, it should come as no surprise that the interior of the human body offers some of the richest habitats available. It's warm and it's wet but it's not so wonderful once the immune system cranks itself up to destroy invaders.

In every moment of every day, every person's body is a battleground where microscopic aggressors constantly seek their biological destiny to multiply and be fruitful. Most of the time, of course, most people remain healthy because their bodies are able to repel the unwanted boarders. The defenders are, of course, the cells of the immune system, the part of the body that, after the brain, is probably the least understood. Unlike the brain or any other organ or tissue, the immune system consists not so much of anatomically linked solid organs (although it does have some—the lymph nodes, for example) but of many different kinds of free-roaming, independent cells. The cells of the immune system are spawned in the bone marrow, but from there they fan out to inhabit every part of the body except the brain.

The system includes two main avenues of defense:

1. One consists of several types of cells that roam the body looking for aliens. They can tell friend from foe by recognizing certain proteins on the surface (each person has his or her own marker molecules, as unique as fingerprints). When these special cells spot anything that is not homemade, they sidle up next to it and emit chemicals that blast holes in the invader's membrane. Or they may simply eat the enemy.

2. The second method involves a different set of cells,

VACCINES: SHEEP IN WOLVES' CLOTHING

Vaccines work because they contain samples of proteins from potential enemy microbes, sometimes even entire corpses of killed enemies. Because they are dead, the corpses can't multiply in the body and cause disease. Or, sometimes, the bugs' bodies are not actually dead but just crippled, again so that they can't hurt anyone. Either way, the immune system sees these, thinks they're for real, and goes through the process of learning to make perfectly tailored antibodies.

Once done, the cells that make these antibodies continue to live for years. If the real enemy comes along, they are prepared to attack with full force immediately. The invader is vanquished before it can cause the first symptom.

HIV: THE MOST DIABOLICAL VIRUS

AIDS is caused by the ultimate germ. All other microbes attack various organs and tissues within the body, and in most cases, the immune system eventually gets the better of them. The AIDS-causing virus, or Human Immunodeficiency Virus (HIV), targets the immune system itself. HIV preferentially infects and kills specialized cells, called "helper T cells," that are the central administrators of the immune system. These cells tell the other cells of the immune system what to do and when.

HIV infection alone, however, is not directly what kills. Disease arises as a result of all the invading microbes that are unopposed by an immune system and thus free to roam the body at will, destroying tissues wherever they please. These so-called opportunistic infections are what kill people with AIDS.

ones that make antibodies. These are protein molecules shaped to fit foreign proteins carried on the surfaces of invading germs. Antibody binding may disable the invader by itself. Even if not, the coating of antibodies makes the alien look quite tasty to other immune cells that will summarily eat it.

One of the great mysteries of the immune system was how it could possibly make enough different kinds of antibodies to be able to bind to any of the millions of different foreign proteins that might come along. There aren't enough human genes to code for that many antibodies. It turns out that the body starts out with only a small range of antibody types. If one of them happens to come close to fitting, however, the immune system causes the cells that made the nearly right antibody to start dividing and producing daughter cells that each make slight variations on the original. In a sense, the cells evolve. Through an elaborate system of relaying messages, the cells that made even-better-fitting antibodies are stimulated to evolve still further. Eventually, the immune system creates cells that can make antibodies with a perfect fit. This takes a few days, and it is why infections sometimes get worse before the immune system finally overcomes them.

AND—AS THEY SAY
IN ADVERTISING—MUCH,
MUCH MORE!

The human body has many other organs and specialized tissues within organs—too numerous to mention and still call this book *Instant Biology*. We didn't mention the magnificent spleen or the hardworking prostate gland (which only some humans possess), and we gave short shrift to such stalwarts as the pancreas, the liver, the kidney, and the many glands. We didn't go into the way red blood cells take up oxygen or how the liver takes toxic chemicals out of the blood and breaks them down.

But you get the idea. Your body, molded during nine months of prenatal development and shaped by all those subsequent years of use and abuse, is a wondrous work of biology, a veritable symphony of physical and biochemical processes. But it is not unique in any of these respects. The human body works pretty much like the bodies of all other mammals. Indeed, the processes of physics and chemistry that sustain its life are the same ones that act throughout the entire living world.

SUMMARY

⏱ The human body is the product of a process of embryonic development that begins with the union of a sperm and an egg.

⏱ In this process, the original founding cells proliferate and change in form and function according to combinations of selected genes. In humans:

a. The first two weeks after fertilization are devoted to formation of the placenta.

b. During the next six weeks, the embryo forms. Groups of cells crawl and slide over one another, using their specialized abilities in various parts of the embryo.

c. By about eight weeks, all the major organ systems and bodily structures have formed and exist in rudimentary form.

d. The fetal stage, in which the basic structures become more complex and larger, lasts the remaining months of the pregnancy.

e. The process is completed at birth. The abilities to walk, talk, and keep the family car out until way past any reasonable hour will not arise until some time later.

⏱ The human body is a complex machine made of many interacting parts, each with a specialized role to play, be it the movement of an arm or the mounting of an attack on invading microbes.

CHAPTER 5

CYCLES OF MATTER AND ENERGY:

IT'LL ALL BE COMING AROUND AGAIN

YOU MUST REMEMBER THIS

Almost all life on Earth is solar powered. Plants capture the energy in the process of photosynthesis and store it in the chemical bonds that hold together molecules of sugar. The stored energy is passed among other organisms through food chains. This way matter and energy are continually recycled among living things. Some energy is lost at each step, but the total in the biosphere is continually replenished as new plants soak up more sunlight.

SUN WORSHIP: THE NATURAL RELIGION

The energy that powers the human body—and the bodies of nearly all other living creatures—originally came from the sun. We're talking not just about the energy we burn to run our muscles but all the energy used in all the metabolic processes within and between all the cells—from the lash of the sperm tail that propelled your father's genes toward your mother's genes to the pulses of electricity that your brain just shuttled from one neuron to another to read this sentence. The sun is even the original source of the energy that simply holds together the atoms that make up the molecules that make up the cells of your body.

The sun does all this inadvertently. It pours forth vast quantities of photons (you can think of photons as particles of light or bundles of energy) in all directions. We on Earth benefit because there are certain molecules on this planet (mainly the green chlorophyll in plants but also certain other molecules in bacteria) that absorb the photons and use their energy to rearrange the atoms taken from molecules of water and carbon dioxide. The energy that was in the photon (actually, it was the photon) is first applied to glue together the atoms to make molecules of ATP, the universal energy-carrier of all cells (Chapter 3). Then ATP's energy is used to make sugar molecules. Finally, the plant cells that do this link several sugar molecules into a chain called a starch and store it. The sugars may also be modified and stored in the all-too-familiar form of another molecule called fat.

Plants use the stored energy to drive their own life

processes. Nearly all other forms of life get their energy by eating plants or eating other things that eat plants.

THERE ARE EXCEPTIONS TO THE RULE . . . .

Close readers will have noticed that "nearly all" living things use energy that first came from the sun. The exceptions are certain kinds of bacteria that feed on hydrogen sulfide, which forms without the aid of the sun. Each molecule of hydrogen sulfide carries useful packets of energy in the bond holding its hydrogen atoms to its sulfur atom.

Hydrogen sulfide (which smells like rotten eggs) occurs in many places on Earth and supports colonies of these bacteria. The substance is especially abundant in the hot, mineral-rich water that spews from deep sea vents, also called hydrothermal vents. Sulfur-eating bacteria thrive around these structures, which are much too deep for sunlight to reach. The bacteria are eaten by organisms that live nearby. Thus, they form the base of food webs that support a variety of animals from worms and crabs to small armies of marine biologists.

PHOTOSYNTHESIS: TRAPPING THE LIGHT FANTASTIC

The green stuff in plants (and a few nonplants) is chlorophyll, which may be the most fundamentally important substance in the living world. Its job is to catch solar energy and make it available to all other living things. (And the best that TV can call it is a "breath freshener"?!) Chlorophyll is the molecule that absorbs photons (which are packets of light energy) and uses the light energy to make ATP, the "universal battery" that all cells use as an energy source (Chapter 3).

ESOTERIC TERMS
(es-ə-'ter-ik tərms)

Atom: The smallest particle of matter that elements—like carbon, hydrogen, and oxygen—come in. Bind two or more atoms together chemically, and you've made a molecule.

Molecule: A particle made of two or more atoms that stick together because of chemical forces between them.

Chlorophyll: The green molecule in plant cells that catches solar energy and traps it in the form of bonds that hold together the atoms that make sugar molecules. This process is called photosynthesis.

Photosynthesis: An elaborate series of chemical reactions in which chlorophyll catches solar energy and converts it to the chemical energy that holds atoms together to make molecules. Ultimately, that energy is used to rearrange the atoms of water and carbon dioxide into sugar, oxygen, and some more water.

Entropy: The scientific term for the degree of randomness or disorder in processes and systems. It is a law of nature that in the absence of outside energy sources, things disintegrate. Without photosynthesis, for example, life would increase in entropy and living things would, in a sense, disorganize themselves to death.

Decay: Life has its own ways of making things disintegrate. Countless organisms feed upon the corpses of once-living organisms. Virtually all decay is a result of other things feasting upon the departed.

Chlorophyll does its work inside chloroplasts, which are organelles (or little organs) in plant cells. Also in the chloroplasts are other molecules that immediately use ATP's energy to carry out a chemical reaction between water (which plants take in through their roots) and carbon dioxide (which the leaves soak up from the air). This chemical reaction produces sugar, which everybody knows has lots of calories, and calories, after all, are a measure of stored energy.

The whole process of photosynthesis is an enormously complex series of chemical reactions, but it can be summarized in one fairly simple chemical equation. You

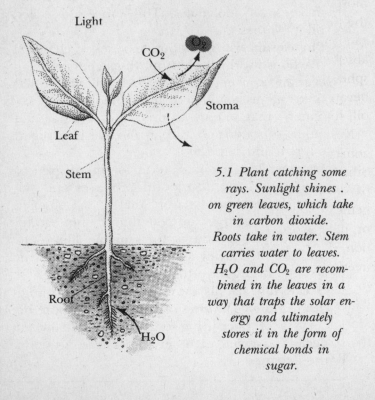

5.1 *Plant catching some rays. Sunlight shines . on green leaves, which take in carbon dioxide. Roots take in water. Stem carries water to leaves. H_2O and CO_2 are recombined in the leaves in a way that traps the solar energy and ultimately stores it in the form of chemical bonds in sugar.*

probably had this one in high school biology, but here
it is again because it's worth looking at:

$6\ CO_2 + 12\ H_2O = C_6H_{12}O_6 + 6\ H_2O + 6\ O_2.$

In English, it says that the atoms in six molecules of
carbon dioxide plus twelve molecules of water are regrouped to yield one molecule of sugar (that's the long
formula—$C_6H_{12}O_6$—which stands for glucose, the most
common of several kinds of sugar) plus two byproducts
that plant leaves simply dump into the air—six molecules
of water and six molecules of oxygen. This is the process,
repeated in innumerable leaves over millions of years,
that has created most of the oxygen in the atmosphere—
millions of billions of tons of it. The process also re-
moves tens of billions of tons of carbon dioxide from
the air every year, for which we should be grateful because carbon dioxide is a "greenhouse gas" (one that
absorbs heat and raises the temperature of the atmosphere). The total concentration of CO_2 has been rising
because of the extra amounts emitted by burning coal,
oil, natural gas, wood, and all other organic materials.

Now you know why sugar is called a carbohydrate—a
combination of carbon and hydrate, which is water. Another carbohydrate is starch, which is simply a polymer
of sugar, a chain of many sugar molecules strung together. Plant cells make starch as a way of storing sugar
or, more explicitly, storing energy and materials (the carbon, hydrogen, and oxygen atoms) for future use. When
needed, the enzymes in plant cells can rework the starch
to make amino acids, the building blocks of protein.
Plants do this, but so can animals that eat the plants.
Most of the plant foods that humans eat are, after all,
the starchy parts—the grains, the tubers, the roots.

Plants can also convert sugar to fat, which is almost

twice as efficient a way to store energy. One gram of fat, for example, contains nine calories, while the same weight of carbohydrate or protein contains just four calories.

So photosynthesis is the fundamental process that sustains virtually all life. It provides both the raw materials

BREAKING THE SECOND LAW

It can't be done, no matter what your crazy uncle claims.

The famous "second law of thermodynamics" says that any time you convert energy from one form to another (sunlight to the chemical energy in ATP, for instance) or extract it from one storage place for use in another (using the energy in sugar to drive a chemical reaction), some of the energy is lost. It escapes as heat. The bottom line: all uses of energy waste some of it. This is a restatement of the concept of entropy—the law of physics that says organizations (which would include living organisms) in closed systems gradually run down, peter out, and fall apart.

Now here's the tricky part: There are people who claim that life must be a miracle because it violates the principle of entropy, that it violates the second law. They say life, which may be the universe's most highly organized system, shouldn't be able to keep on going eon after eon unless some supernatural force were sustaining it.

The problem with this argument is that life is not a closed system. Photosynthesis is always pumping new energy into the system. Switch off photosynthesis, and in a couple of months virtually all life would cease on Earth.

and the energy that plants need to grow and, therefore, supports the rest of the living world, which feeds on plants, directly or indirectly.

FOOD CHAINS: WE DON'T MEAN MCDONALD'S

Fast-food franchises are not the kind of food chain we're talking about here. But, come to think of it, McDonald's—and all other places where humans eat—does occupy a position in the food chain of which our species is a part: Grass grows by photosynthesis. Cattle eat grass. Humans eat cattle. That's one example of a simple food chain.

Sun → grass → cattle → humans.

But, of course, this is not the only food chain to which we belong. Along with the all-beef patty, there's lettuce, pickles, tomato, onion, and special sauce, not to mention the sesame-seed bun. So we, like most species, belong to several food chains. We really belong to a food web.

If you go for the fish sandwich instead, the food chains are more interesting. They start with algae, which grow by photosynthesis. Tiny animals that make up the plankton (critters so small they drift in the water more than swim in it) eat algae. Little fish eat the tiny planktonic animals. Big fish eat little fish. We eat big fish. And, of course, the lettuce and tartar sauce make it a food web. Order the fries and you're solidly linked into several of Earth's grand cycles of relaying matter and energy among various life forms.

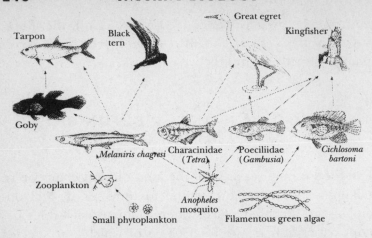

5.2 It's a life-eat-life world. One small example of a food web. Arrows show who feeds whom.

Food webs in the wild can be more complex. In some wetlands, for example, the marsh grasses are eaten by small grasshoppers. Orb-weaving spiders trap and eat the grasshoppers. Warblers eat the spiders. Marsh hawks eat the warblers.

Other than organisms that can perform photosynthesis, all living things depend on a food chain of some sort.

GOODNESS VORACIOUS

Biologists group species into "trophic levels" (which are levels in a food chain) according to where they get their energy. The familiar categories of plant eaters (herbivores) and meat eaters (carnivores) are a bit of an oversimplification. Here's a more complete rundown of who's coming for dinner.

GOODNESS VORACIOUS (*continued*)

Trophic Level	Examples	Source of Energy
Photosynthesizers	plants, some monerans, and protists	solar energy
Herbivores	grasshoppers, anchovies, ducks, cattle, vegans	plants
Primary Carnivores	spiders, wolves, warblers	herbivores
Secondary Carnivores	tuna, hawks, dolphins	primary carnivores
Omnivores	bears, crabs, opossums, most humans	all of the above
Detritivores*	fungi, worms, many bacteria	dead remains and waste products of all categories

*Detritus: dead and wasted organic matter

ROT. WITHOUT IT, WE'D ALL BE DEAD

Food chains and food webs are one-way paths in the biological commerce of matter and energy. To have a cycle, there needs to be some way of returning these commodities back to the beginning. That's the job of organisms that carry out decomposition—the agents of rot.

You wouldn't know it from the average teenager's bedroom but nature is full of organisms that quickly clean up the remains of dead and discarded stuff. Actually, if you look closer, you can see it. That pizza crust under the bed with the green fuzzy stuff on it? That's a food chain trying to circle back to its beginning. If it weren't for civilization, the wheat in that pizza crust would be on the ground somewhere and, as it rotted, the energy and materials in it would sink into the soil and enrich it to help new plants grow. Left to its own devices, nature composts everything.

So when things decay, it's not a sign of something gone wrong—not in nature's grand scheme of things. It's a sign that nature is reclaiming energy and materials that seem to be no longer needed by higher organisms. The worms and insects that feed on the larger fragments from animal wastes and dead bodies, the fungi and bacteria that break down the smaller stuff, these decomposers, or detritivores, are life's ultimate recyclers.

Animals that aren't eaten by other animals can't have their matter and energy recycled without rot. And unless almost all organic matter is recycled, the next generation of plants will be deprived of nutrients. And if the plants don't grow, the animals won't eat so well. It's a process that has been going on for millions of years. In other words, you are what the dinosaurs ate.

FOSSIL FUELS:
ANCIENT SOLAR POWER

Over much of life's history, the amounts of matter and energy being cycled over the planet have stayed fairly consistent. The new growth put on by the world's veg-esphere (that's a made-up word, so don't try it on a biology professor) is pretty evenly cropped by the animals. And the decomposers return enough formerly living matter and energy to the soil to keep the plants going. Everything works out (where industrial development has not intruded). But it was not always this way.

There was a time when the plant kingdom outpaced everything else on land. So many plants lived and died without being consumed by other organisms that their remains—still rich in organic matter and its stored energy—exist today, altered only by chemical and physical decomposition. They are the world's vast coal fields. By the same token, epochal blooms of marine organisms died but were not rotted by decay organisms. They are today's underground oil pools.

The luxury of the modern era, energized by fossil fuels, was made possible by the way life evolved on Earth.

This age of plants happened around 340 million years ago, before the animal kingdom became as diversified on land as it is now. The plant kingdom grew luxuriantly over a largely tropical planet. Vast regions of ferns, moss, and other primitive greenery covered the landscape, many of the plants growing into tree-sized monsters. These were the "coal forests," as biologists call them. After layer upon layer of dead vegetation piled up and the whole mess was eventually covered by sediments, the stuff was squeezed and transformed into a black rock rich in organic compounds. What once was sugar made by photosynthesis and then rearranged into starches and tougher carbohydrates like cellulose (the main carbohydrate molecule of which wood is made), all this fossilized plant matter is now coal. Much the same story, only involving marine plankton, is what gave us natural gas and oil.

The solar energy captured by photosynthesis is what

we release today when we burn fossil fuels. The heat from the fireplace, whether you burn wood or gas, was generated in the bowels of the sun. And the carbon dioxide that is released as a by-product, with all its global-warming potential, is essentially the gas that was removed from the atmosphere when ancient forests luxuriated in a very much warmer world.

PRODUCE DEPARTMENTS

Habitats differ hugely in the amount of life they can support, mainly on account of the amount of rainfall but also because of temperature differences. The tropics are the world leader in sheer quantity of plant matter per square mile. They cover just 10.8 percent of the Earth's surface but harbor a whopping 56.1 percent of its plant biomass. Here's how the regions stack up:

Vegetation zone	Percent of world's area	Percent of world's plant mass
Polar	1.6	0.6
Conifer forest	4.5	18.3
Temperate	4.5	11.5
Subtropical	4.8	13.5
Tropical	10.8	56.1
(Total land	26.2	100)
Glaciers	2.7	0.0
Lakes & rivers	0.4	less than 0.01
Ocean	70.7	less than 0.001
(Total water	73.8	less than 0.01)
(Total Earth	100	100)

SUMMARY

⏱ All life on Earth participates in one vast network of cycles that take the matter and energy of each organism (some only when they are dead) and recycle those commodities to give life to others. There are many different food chains, and many of them are interlinked, creating one vast, global "food web."

⏱ Solar energy, which begins the cycle, drives photosynthesis in green plants, turning water and carbon dioxide into sugar (which captures and holds some of the solar energy) plus two waste products, water and oxygen.

⏱ Plant cells turn the sugar into organic compounds such as starches, fats and proteins to build plant tissue. Animals eat plants or they eat other animals that have eaten plants. Either way, the matter and energy is passed along the food chain to the "top predator."

⏱ The final stage is the process of decay—the ultimate recycler. Insects, worms, bacteria, fungi and other organisms work on the corpses of dead animals and plants, dismantling their tissues and cells. This raw material returns to the beginning of the food chain to enrich the soil to fertilize new plants.

CHAPTER 6

THE INTER-DEPENDENCE OF LIFE:

WE'RE ALL IN THIS TOGETHER

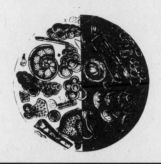

YOU MUST REMEMBER THIS

Each species of living thing belongs to a community of many species, and those communities exist only where the physical conditions (such as temperature, rainfall, and soil type) are right for that community. Within each community, organisms may cooperate or compete for resources, eat or be eaten. Whatever the arrangements, each species depends on one or more of the others for survival.

FORGET IT, GRETA.
WE DON'T *VANT* TO BE ALONE!

Nowhere on the planet can you find any species of plant or animal that lives by itself. It's not that all species are so gregarious (some are, some are not). It's that there is a built-in biological necessity to depend in some way on other species. No species *can* live alone, not even the least sociable microbe on the sea floor. One important and rather intimate type of relationship among species was cited in Chapter 5: some of us eat others of us. Some animals, for example, eat plants. Other animals eat those animals. And in the end, the species of the microbial world—mainly the bacteria and fungi—eat us all, even if we escape the clutches of larger critters.

But the dietary link is far from being the only one that binds species together. There are many other links. Think, for example, about the plants that depend on bees and other insects to mediate their sexual liaisons. Most flowering plants can't get themselves pollinated (fertilized by pollen, which are male sex cells from another plant or another part of the same plant) unless they bribe insects with nectar. There are many examples of plant species that can be pollinated by only one species of insect. And for that insect species, the only source of food is that one kind of plant. Neither species can exist in a place that lacks the other. There are many other examples of plants that can make use of several kinds of insects and these insects visit several kinds of flowering plants.

But in no case can a flowering plant grow just anywhere. Plants are pickier than animals. Each species needs a certain type of climate. The seasons have to provide just the right temperature ranges. The rainfall must

ESOTERIC TERMS
(es-ə-'ter-ik tərms)

Ecology: The science that studies relationships among living things and their environment. The word comes from the Greek *oikos,* meaning "house," and *logos,* meaning "study of." Ecology does not mean "the environment."

Habitat: The type of environment in which a particular species lives. The bullfrog's habitat, for example, is the edge of a pond. The blue whale's habitat is, well, all the ocean he wants.

Community: All the living things that dwell in a particular habitat. Because they interact with one another, the various species in that habitat are thought of as a community.

Ecosystem: The organisms living in a particular habitat (the community) plus the physical part of their environment, which includes the soil, the water, the climate, and everything else in it.

Biome: A really huge ecosystem, such as the grasslands of East Africa or the tropical forest of South America.

Ecological niche: A way of making a living (or exploiting a resource) that either is being used by some species in an ecosystem, or could be. For example, woodpeckers occupy the ecological niche that makes use of dead, standing trees, both for nesting holes and to find food (insects that live under decaying bark).

be adequate, but not so much that its roots become wa-
terlogged. And each species of plant needs soil of a certain mineral composition with a certain amount of
organic matter. Then there's sunlight. Some plants
thrive only in full sun, others need shadier environs. Because of all this choosiness, each kind of plant can live
only in certain parts of the world and only in certain
locales in those places. And so its pollinators must live
in those same places.

Fortunately, there are many kinds of plants and they

LOCATION, LOCATION, AND LOCATION

The major factors that determine the ecological value
of any piece of real estate are physical characteristics
that depend largely on where on Earth the place is.
Depending on location, several physical features define
local environments and determine what kinds of plants
can live there. Then, to a large extent, the plant community determines what kind of animals live there, or
would, if they could afford the airfare.

Here are some of the main factors that species must
consider before deciding whether to buy in any particular neighborhood.

Climate:

Temperature—Day-night variation and annual
seasonality

Seasonality—Cold-warm cycles and/or rainy-dry
cycles

Day length—Nearly constant in tropics but varies
seasonally toward poles

Rainfall—Total amount and its distribution over a
year's time

have very different environmental preferences. Some, for example, prefer swampy land while others like sandy deserts; some live only in warm, tropical conditions while others will die unless they get severe winters every year. This means there are plants of one sort or another that will grow just about anywhere—except, of course, the grass in your own yard.

Species are interdependent in other ways as well. Birds that nest in cavities in trees, for example, would be home-

LOCATION, LOCATION, AND LOCATION (*continued*)

Soil type:

Particle size—Rocks, gravel, sand, clay

Organic content—The more, the better most plants like it

Chemical composition—Different combinations of minerals; acidity vs. alkalinity

Water:

Fresh water—Soil saturated or flooded, periodically or all year-round, flowing or standing

Salt water—Degree of salinity, depth, and temperature of water (near shore to deep sea), available sunlight, dissolved and suspended nutrients

Altitude:

Higher—Cooler temperatures, more intense sunlight, stronger winds

Lower—Generally warmer, less intense sunlight

Disturbance:

Nonhuman—Storms, earthquakes, landslides

Human—Pollution, mining, hunting, fishing, other people's headphones that only let you hear the hissing, clashing rhythm

less without trees. Some would still be homeless if it weren't for woodpeckers. The woodies can excavate their own holes, but some other birds depend on these hard-headed avians for the construction work. They move in after the 'peckers move out. Cottontails, as Brer Rabbit knew, can find safety in briar patches that protect against predators. But the cottontail's desert cousins, the jackrabbits, can survive in open country because they have bigger ears (the better to hear danger approaching) and more powerful jumping legs (the better to escape). And the white-tailed deer prefers to dine on the leaves of shrubby plants and trees small enough that the deer can reach the leaves. This is why you're more likely to see deer along the roadway than deep in the woods. A mature forest doesn't offer much tasty food at deer level, but the forest edge, where sunlight reaches the ground, does.

The point of these examples is that plants and animals live together in relatively predictable combinations. You won't find jackrabbits in the forest or deer on the open prairie.

HOME IS WHERE THE HABITAT IS

The major types of habitats—or biomes, if you prefer—on planet Earth are these:

Water-based ecosystems vary from the largest, wettest places (the oceans) to those where mere trickles of wet stuff pass through every now and then (desert streams).

Oceans

This is by far the biggest biome, covering 70 percent of Earth's surface, but it is not evenly populated with

6.1 Seashells by the seashore and much more: An ocean habitat.

life. The density and diversity of living things are greatest along the coasts, lowest in the open ocean. The biologically richest marine habitats are estuaries—places where fingers of the oceans reach inland to join with shallower waters that receive much organic matter from the surrounding land areas. Chesapeake Bay, for example, is an estuary and one of the world's richest marine habitats, producing millions of pounds of crabs, oysters, and fish a year. A key feature of estuarine ecosystems is the amount of salt in the water. Near the open ocean, salinity matches that of the sea. But as one moves toward the freshwater rivers that feed most estuaries, the salinity declines gradually. This gradual change is matched by a series of changes in the communities inhabiting the water.

In the open ocean, habitats vary according to depth and latitude. In general, the shallower, the warmer the water, the greater the variety. Coastal waters support rich communities of plants, animals, and microbial life. In tropical waters, coral reefs (the built-up skeletons of coral animals) are the most biologically diverse ocean communities. With greater depth, the richness declines, mainly because less sunlight penetrates. In the deepest ocean, perpetual darkness reigns, but still there are communities adapted to life there, feeding on the organic matter that sinks from the sunnier realms above.

Freshwater Lakes

These vary from huge lakes, such as the Great Lakes (which are almost like oceans) to lesser lakes, such as neighborhood ponds. These are usually rich in aquatic plants and insect larvae, along with fish, frogs, and even some reptiles, such as turtles and water snakes. The smaller the lake, the faster it undergoes a natural process in which

6.2 On golden ponds and lakes: A freshwater aquatic habitat.

the bottom gradually fills in with sediment and the water becomes richer in organic compounds. This process, called eutrophication, can be speeded up manyfold if people drain nutrient-rich sewage or farm runoff into the lake. As eutrophication proceeds, the communities inhabiting the water gradually change. Cold, deep clearwater lakes populated by trout, salmon, whitefish, and commercial fishermen gradually become shallow, algae-choked ponds with bluegills, bass, and boys with cane poles.

Streams

Where it rains, it pours. Water that is not immediately absorbed by the soil simply flows downhill, seeking its lowest level. Most of the life that depends on stream flow lives along the water's edge. In some deserts (where little rain falls anyway), the only plant life may be along the stream (the only place enough water collects to support

6.3 Down by the old freshwater stream habitat.

plants). Depending on the speed of the flow, stream habitats may host communities of insects, fish, crustaceans (such as crayfish), frogs, snakes, turtles, and, of course, fly-fishermen doing their best to mimic insects that flit over the water, tempting the trout just below.

Land-based ecosystems range from the mostly frozen realms near the poles through the ill-named temperate zones (what's so temperate about fluctuating between freezing and sweltering?) to the perpetually warm and often wet tropics of the equatorial regions. Habitat types may vary within the polar-equatorial gradient depending on a number of factors. Altitude is a major one; for every 1,000 feet of altitude, the average temperature can be 3 to 4 degrees Fahrenheit cooler than below. As a result, a mountain habitat at one latitude may resemble the sea-level habitat far to the north of it (in the Northern Hemisphere, where people stand right side up). Among the broad categories are those that follow.

Tundras

In the regions that are coldest year-round, the climate is too frigid for nearly all but the hardiest small plants that can get by on growing seasons of only a few weeks or months. The highs are typically only in the 30s. No tree species can stand this, so the northernmost reach of trees—the timberline, or tree line—often defines the southern limit of the tundra. The chief plants are moss, lichens, grass, and dwarf shrubs. The year-round vertebrate community is often dominated by musk oxen and polar bears. In warmer months, the tundra is visited by caribou, snowy owls, ptarmigans, and occasional ducks and geese. Though limited in big species, the tundra has

6.4 A land where the tall trees don't grow: a tundra habitat.

a respectable complement of insects and smaller creatures, especially near its many ponds and bogs.

Forests

Though among the more familiar ecosystems, forests are, in fact, highly complex and poorly understood. They vary enormously in type, from the all-conifer forests that blanket northern latitudes just below the tundra to the amazingly diversified tropical forests, which, though relatively small in area, are home to more than half the planet's biodiversity. Where a northern forest may have just one or two species of tree mile after mile, a tropical forest may have several hundred different tree species in just one square mile.

In temperate climates there are two broad types of forest: coniferous (evergreens, such as pines, hemlock, and spruce) and deciduous (hardwood species that lose

6.5 This is not the forest primeval, but even as second-growth, it counts as a rich forest habitat.

their leaves each fall, such as beech, oak, and maple). The largest such forests are in Asia and North America.

In the tropics, there is little temperature change, and most trees have green leaves all year-round, shedding old ones and growing new ones at the same time. Examples of these steady-state trees are palms, mahoganies, and the thousands of other species that make tropical forests such havens of biodiversity. In addition to tall trees, which place most of their leaves high in the air, tropical forests usually have dense plant communities in the shadier regions below the trees' canopy. These include vines and plants, called epiphytes, that grow on the branches of woody species, creating the habitat popularly known as jungle. As a result, tropical forests have very different plant and animal communities that live on the ground or in the uppermost canopy or somewhere in between. The largest tropical forests are in the Amazon River basin of South America and the Congo River basin of Africa.

Grasslands

These are the plains, prairies, and savannas that generally occupy the interiors of continents. In many cases, they would be forests except that there are prolonged

6.6 Amber waves of grassland habitat.

dry seasons that make life too hard for trees. Grasses are well adapted to drought because they have vast root systems that can reach down many feet to tap soil moisture. Even if the green part of the grass dies, grasses can remain alive just below the surface for many months. This is why grasslands bounce back so quickly after fires, which people have been setting in grasslands for thousands of years—usually to encourage the tender, tasty new growth that draws plains animals (the better to be hunted) or that fattens domestic livestock (which you don't have to hunt but do have to feed).

In North American grasslands (at least in the remnants that haven't been put to the plow for the growing of domesticated grasses, such as wheat and corn), as many as seventy wild grass species may cohabit in the same area. Grass is such a popular food for animals that grasslands support huge amounts of animal life. East Africa's Serengeti Plain is one example, with its wildebeest, many other kinds of antelope, zebra, giraffe, elephant, rhinoceros, lion, cheetah, hyena, and other stars of *National Geographic* specials.

North America's grasslands were once as abundant, until the arrival of European settlers, steel plows, and the railroad. The prairie was home to such companions as pronghorn antelope, bison, and elk, along with their predators, such as wolves, coyotes, and cougars. Today, a few remants of these large species share the grasslands with badgers, prairie dogs, and rabbits. There are also numerous birds, such as larks and plovers and predatory hawks and owls. Where rainfall can sustain marshes and flood small depressions called potholes (they were out on the prairie before they showed up in city streets), the grasslands provide essential habitats for numerous ducks, geese, coots, and other waterfowl.

6.7 A desert habitat, but definitely not deserted.

Deserts

In any latitude or altitude where the rainfall is slight (generally less than ten inches a year), you have a desert. There are several kinds of desert, including the sand dunes of the Sahara in Africa, the cactus- and roadrunner-rich Sonoran Desert of Arizona, and the dry valleys of Antarctica where almost nothing lives. Even in sand dunes, however, many small creatures, such as insects and lizards, live, spending their days in the sand, out of the sun, and emerging only at night. Desert plants survive because they have evolved various ways of coping with dry conditions. Cacti, the classic example, have expanded their stems to hold large amounts of water, reduced the surface area of their leaves (which would otherwise give off water), and converted these anatomical organs to thorns that protect the stem from animals. In place of the leaves, the bulging stems are the green sites of photosynthesis.

GUESS WHO'S FOR DINNER

Within any habitat, the most intimate relationship one species can have with another is to eat it or, if not so lucky, to be eaten by it. In fact, ecologists have concluded that no matter what the ecosystem, there are only a few major ways—such as this predator-prey relationship—in which species interact.

Other kinds of relationships come under the category of symbiosis. The word is from the Greek *sym*, "together," and *bios*, "life," and it refers to various kinds of lasting relationships between individuals of different species.

Here's a list of relationships, starting with the one that is not symbiosis:

Predator-prey

Thanks to innumerable nature shows, this is probably the best known. Lion sees wildebeest. Lion chases wildebeest. Lion whaps ungainly 'beest with mighty paw and dinner is ready. What usually doesn't make it in teeveeland documentaries are all the other predatory-prey relationships that happen right in your own backyard and, indeed, even in many people's houses. The robin taking the worm is a predator catching prey. And that spider-web up by the living room ceiling? That's no reflection on your housekeeping; it's a patch of nature indoors, a predator's trap for unwary moths.

Predator-prey relationships are famous because they are the driving force behind the evolution of many unusual features in the plant and animal worlds. Critters at risk of being eaten gain a major survival advantage if they happen to evolve a defense mechanism that lowers their odds of being caught. Thus, any change in coloration that might confer camouflage is a big help, as would be an improved ability to run away or to fight back. Some prey species have spectacular defense mechanisms, such as poison glands or the skunk's repellent spray. Other species find refuge in looking like something else. That's why katydids look so much like green leaves and fawns have spots that look like the dappled light where their mothers hide them.

LINKS BETWEEN LYNX AND HARE THERE (IN CANADA)

One of the classic studies of predator-prey relationships shows how Canadian lynxes and snowshoe hares each control the population size of the other. The discovery came from studies of the numbers of pelts of each species that trappers were selling on the fur market each year. The number of white hare pelts would soar for a few years, then crash to almost nothing. The number of

LINKS BETWEEN LYNX AND HARE THERE (IN CANADA) (*continued*)

spotted lynx pelts would climb, peak, and also crash. But the two cycles—each varying from nine to eleven years—were not in exact sync. The hares peaked before the lynxes and rebounded before the lynxes.

Ecologists figured out what was going on. The hare's chief predator is the lynx. When the hare population crashed, there was less food for lynxes, so their numbers fell, too. As the number of lynxes dropped, the hares could—as hares do—multiply rapidly once again. But, as hare numbers rose, the lynxes' food supply did, too, and lynxes made a comeback. Because lynxes have longer life cycles than do hares, their numbers climb more slowly. Once they've gobbled up lots of hares, sending the bunnies into decline, the longer-lived lynxes were similarly slower to follow with their population crash.

Host-parasite

This is the only other relationship, along with predator-prey, in which one member definitely gets the better deal. In this case, one organism lives in close, permanent association with another and derives a benefit, while the host suffers. For example, a tapeworm living in the human gut is a parasite, stealing nutrients that would otherwise benefit the person. Bacteria that infect living animals (including people) are also parasites, deriving their sustenance from the host. Even plants can be parasites. Mistletoe, for example, parasitizes trees, sending its rootlike structures into the tree's branches to get water.

Mutualism

Sometimes two species cast their lot together and work out a deal that is good for both. In fact, many mutualisms are essential to the survival of both species. The best known example is the lichen—that flattish, crusty growth that forms on rocks and tree trunks. A lichen is a partnership between a fungus and a one-celled alga (members not just of two different species but two different kingdoms). The alga performs photosynthesis and supplies the fungus with the sugar it produces, as well as other nutrients. The fungus protects its green friend and provides moisture it absorbs from the air.

Human beings, like most animals, are deep into a form of mutualistic relationship with the bacteria that inhabit their guts. We give the little bugs a home (we are their ecosystem) and all the food they can eat. They digest stuff in our food that we can't. We get the last laugh, though, because then we digest some of the bugs and get the nutrients back. It's still mutualism because the bacterial colony continues to thrive.

Commensalism

When one partner gets a benefit but the other couldn't care less, ecologists call it commensalism. The mites that live in the follicles of our eyelashes enjoy the benefits of this deal. They get a life; we usually never notice.

A more visible commensal relationship in the plant kingdom involves species like bromeliads, the pineapple relatives that live on the branches of trees in the tropics. Bromeliads don't need soil nutrients, so they have no roots. Instead, they have tendrils that grab onto tree branches just to hold themselves in place up off the ground. That's all they get from the tree. Bromeliads get their nutrients by trapping water and falling organic matter in cups formed by their flowerlike leaf clusters.

YOU OTTER REMEMBER THIS

Some ecosystems are held together by just one species whose existence is crucial to the survival of the whole system. Take the case of the sea otter and the submerged forests of giant kelp that once dominated the ocean off the west coast of North America. The kelp grew densely, reaching its huge, fringed, fleshy leaves from anchors on the bottom to the water's surface. The forest was home to a community of fish, crabs, squid, starfish, mollusks, anemones, and sea otters. It was as rich and diverse a habitat as any temperate zone forest on land.

But otters have fine, silky fur that is good for making hats and other clothing. So by the early twentieth century, the mammals had been hunted almost to extinction. Funny thing was, as the otters disappeared, so did the kelp forests and the whole community that lives in

YOU OTTER REMEMBER THIS (*continued*)

them. Turned out that one of the otter's chief foods was sea urchins. With no otters, the sea urchin population exploded. And urchins eat kelp. Once they cleaned out the kelp, the urchins nearly vanished, too.

When conservationists imported a few sea otters from remnant populations off Alaska and southern California and established them at various locales along the Pacific coast, the animals found enough sea urchins to get by and otter numbers gradually started growing. With the urchin population low, the kelp came back, spreading in from other areas. As the underwater forests grew up again, all the other species reappeared, too. Because of this effect, ecologists call sea otters a "keystone species," drawing an analogy between them and the wedge-shaped stone that masons place at the top of an arch to keep the whole thing stable.

THE CAT AND THE CLOVER

Some ecological relationships are not obvious. For example, the more cats there are on a farm, which is its own habitat, the more clover in the pasture. How? Cats kill mice in the fields. Mice kill and eat bumblebees, digging into their underground nests. Bumblebees are prime pollinators of clover.

So, if the cats keep the mice down, more bumblebees will live to pollinate clover. And thus the clover will make more seed and spread itself more thickly in the pasture.

LIFE'S A COMPETITION, THEN YOU DIE

One of the hallmarks of life in any community, including ecological ones, is competition. Different species struggle against one another for access to the same set of resources. The more marginal the habitat (the scarcer the resources), the stiffer the competition. Even members of the same species compete against one another. For example, the roots of plants may intertwine in the soil, each trying to soak up water that the other would like to have. Birds compete for a limited supply of insects or seeds. Predators compete for prey. Trees in a dense forest compete for sunlight.

As in predator-prey relationships, this competition favors the survival of individuals and species with advantages. The bird with better eyesight to spot a camouflaged insect is going to outcompete the bird with poorer eyes. Under these conditions, myopic birds would decline in numbers (as the hawk-eyed birds proliferate) or even disappear entirely. Or it might look for another line of work—maybe switch over to eating plants, which are easier to spot or, perhaps, get a night job that makes better use of its keen sense of hearing, if it happened to have one.

It is competitions such as these that determine the makeup of ecological communities. Each way of making a living—each ecological niche—is an opportunity waiting to be exploited by an existing species or by a new species that might evolve in the future.

Detailed ecological studies have found that even though a given community may host seemingly similar species or different species seemingly competing for the same resource, there are actually significant differences that allow them to coexist peacefully. For example, mice and ants in Arizona's Sonoran Desert were both found to feed on seeds but, on closer inspection, it turned out that there was a difference in the preferred sizes of seeds. The ants, as you might guess, went predominantly for smaller seeds. By preferring different sizes, these two species appeared to be minimizing their competition and each made a decent living.

On taking a closer look, however, scientists discovered that ants sometimes gathered seeds of the size preferred by mice, and mice sometimes took seeds of the sizes that ants liked. This suggested to ecologists that maybe there was a competition after all, so they tried an experiment. They removed ants from some sites and mice from others and watched what happened. Population densities of the remaining species increased. Clearly, mice and ants were in competition for the same resource. Still by specializing in slightly different ecological niches, the two species were able to coexist on the same food sources, with neither species starving out the other.

Species also coexist by occupying different parts of the ecosystem or occupying the same parts at different times. Some species, for example, are out and about only at night, when rivals for some resource are probably asleep. Then, come daybreak, the two species trade shifts. Tropical rainforests are compartmentalized vertically, some species spending their entire lives in the topmost canopy, others on the ground.

DONNA REED'S KITCHEN

It's probably the only pristine environment in the world, the only place where everything is just so, and stays that way. Fiction aside, there is, in fact, no such thing as a pristine habitat. Ecosystems are not fixed entities; they are dynamic, constantly changing webs of interrelationships among many species. As in the story of the Canadian lynx and the snowshoe hare (see sidebar above), relationships among species cause the numbers of each to shift continually. The glory of the natural environment is not that everything is—or was—perfect and fixed; it is that innumerable species are constantly interacting among themselves and with the physical environment (which also changes) in ways that produce a wondrous, kaleidoscopically shifting panorama.

In many ecosystems, however, there is a stability of a rough sort—a broad range of ecological possibilities through which the system cycles over time. The changes are sometimes too slow to be noticed in a human lifetime, sometimes not. One example of a very dramatic change in a short time is caused by beavers. They cut down trees, dam a stream, flood a valley, and drown hundreds of trees with their newly made pond. Over a few years, the pond acquires fish, frogs, cattails, and other inhabitants and becomes a proper aquatic habitat where once there was only a forest stream.

Then, as it does to all lakes, eutrophication happens. Sediments carried into the beaver pond accumulate on the bottom, building it up. Nutrients wash in from the surrounding land, encouraging algae and other aquatic life, which speeds eutrophication. Gradually the pond bottom rises and, because the edge is shallowest, the pond gets smaller. Eventually, the pond becomes a marsh, then a bog and, finally, a meadow. The beavers move on, and grasses start growing on the rich soils created as the pond bottom rose.

Now begins a process called *ecological succession*. It can follow any major disturbance to an area, such as fire, flood, plowing, or even strip-mining. All it takes is time, which may vary from a few decades (in a well-watered environment, such as the eastern United States or the Amazon rainforest) to thousands of years (say, in arid habitats blanketed by volcanic lava and ash).

When a habitat is destroyed, the place can't immediately go back to being the way it was. In a sense, it has to start small and build up again, often over decades or centuries. Only certain species are capable of taking up residence in a disturbed environment. Ecologists call them pioneer species. They invade first and prepare the way for others. When a farm field in the eastern United States is abandoned, for example, grasses and wild herbaceous (nonwoody) plants are the first to colonize the site. Then come small shrubs and, eventually, pioneer species of trees, usually softwoods. Hardwoods are the last, forming a mature forest. The end of the line in ecological succession is called the climax community. In this example, it would be the hardwood forest.

Ecological succession has always worked before—except, of course, when some change or other caused the extinction of a species—but researchers now wonder whether it will be able to cope with the most massive, widespread environmental destruction the world has seen since the mass extinction that wiped out the dinosaurs sixty-five million years ago. The cause this time is, of course, human rapaciousness.

SUMMARY

⏱ All living things are linked to one another in a complex web of relationships. Each form of plant, animal, and even microbial species depends on one or more other species for necessities such as food and shelter.

⏱ The exact combinations of species that live together in communities or habitats are dictated by the physical features of any given area: its temperature range, rainfall, soil type, etc.

⏱ Within each community, species often compete for access to resources such as food, water, or housing sites. The need to succeed under these conditions means that most species either possess anatomical or behavioral specializations that give them an edge over the competition or are adapted to ecological niches so narrow that they do not compete with anyone else.

CHAPTER 7

EVOLUTION:
THE RESULTS ARE ALMOST MIRACULOUS

YOU MUST REMEMBER THIS

There is no debate among working biologists that evolution is a fact of life. It really happened. It is still happening. Your fundamentalist friends may not buy it, but theologians for most of the world's major religions see no conflict between evolution and religious faith.

AND THIS, TOO

Natural selection, the main process that makes evolution happen, is *not* a random phenomenon. It unerringly selects and encourages changes in species that enhance their survival in a competitive natural world and it ruthlessly destroys changes that do not.

EAGLE OF THE *BEAGLE*

Charles Darwin was the first "creation scientist." The father of evolutionary biology, the man once described as "the Newton of biology" and as "the most dangerous man in England," the man religious fundamentalists have vilified for more than a century, Darwin actually began his scientific career intending to demonstrate the glory of God's handiwork.

7.1 Charles Darwin, evolutionary revolutionary.

"I did not in the least doubt the literal truth of the Bible," Darwin said of his beliefs as he set off, at the age of 22, on his epic, five-year voyage aboard HMS *Beagle*. The *Beagle* was a surveying ship, under command of the deeply religious Captain FitzRoy, who had picked Darwin to be the ship's naturalist not because of his scientific acumen (Charlie was then thinking of a career as a country parson) but because FitzRoy, a stuffy sort who didn't mix with ordinary seamen, wanted a gentleman along for conversation. Unfortunately, Darwin was not so good at this: during almost all his time aboard, he was seasick.

Naturally, Darwin sought as much time on land as possible. He explored the wet jungles of South America and the dry rocks of the Galápagos Islands, looking, collecting specimens, taking notes, and trying to make sense of it all. The voyage gave rise to doubts about the literal truth of Genesis and led eventually to the ideas that propelled Darwin to scientific immortality.

THE RELUCTANT REVOLUTIONARY

Contrary to common belief, however, Darwin did not develop his theory of evolution during the voyage. He came to his ideas slowly, and to publishing them almost not at all. It was six years after the naturalist returned to England (he would never again leave) before he wrote a brief, thirty-five-page outline of his ideas. But he didn't want to publish it. For all the daring in his thinking, Darwin was a cautious and deliberate man. He told his friends he wanted to amass more data before going public. Two years later he expanded the text to 230 pages, but still he was afraid to take the leap.

ESOTERIC TERMS
(es-ə-'ter-ik tərms)

Fossil: One of the most powerful forms of evidence that evolution has happened. A fossil is not necessarily just an ancient bone that has turned to stone. It can be any trace of ancient life, such as a fish scale or a footprint. Some fossil bones have turned to stone (become petrified), but others haven't. They're still made of bone.

Gene: A segment of a molecule (made of the chemical DNA) that carries the instructions telling a cell how to make one kind of protein. That protein, along with thousands of others, helps determine the form and function of the cell. In turn, many cells (which all normally have the same gene) dictate the form and function of the individual.

ESOTERIC TERMS
(es-ə-'ter-ik tərms) (*continued*)

Mutation: A change in the instructions encoded in a gene, which then leads to some change in the individual inheriting the gene. This is one of the two main phenomena that make evolution happen.

Natural selection: This is the other main phenomenon that makes evolution happen. After mutations change the individual organism, the struggle to compete in the natural environment determines whether the change is beneficial. If so, the mutated individual is more likely to survive to reproduce and pass on the new trait.

"I think you two should hit it off—you have 98%
of your DNA in common."

Then in 1856 Darwin looked back and something was gaining on him. A young, upstart naturalist named Alfred Russel Wallace was independently coming to the same conclusions Darwin harbored. Darwin heard about it on the scientific grapevine, and his scientist friends warned that if he didn't publish soon, Wallace would steal his thunder. So Charlie set to work on his magnum opus, which he envisioned as a much-expanded version of the 230-page "outline."

Two years into that project, Darwin was shaken to receive in the mail a crisply written 4,000-word exposition of the theory of natural selection, written by Wallace. "All of my originality," Darwin said, "seemed smashed— and after twenty years of marshalling evidence." Not wanting to see their friend slighted, Darwin's naturalist cronies thought of a way to make the Darwin-Wallace race a tie. They dug up one of Darwin's old unpublished papers and arranged to have it read along with Wallace's at the same meeting of the prestigious Linnaean Society.

A fire under his chair, Darwin finally picked up his pace. He abandoned his plans for the "big book," as he called it, and reduced the project to a 490-page tome, which he considered a mere abstract. *On the Origin of Species* was published in 1859, and all 1,250 copies sold out the first day. Although the church attacked Darwin viciously at first, the theological case against evolution was largely withdrawn during Darwin's lifetime. He died knowing that the major Christian denominations had declared it possible to be a Christian and a Darwinist at the same time.

The rivalry with Wallace also quickly faded. Each man acknowledged the work of the other, and Alfred even graciously acknowledged that Darwin was not only first but had presented a far more elaborate case for the theory. (Such courtly behavior among researchers has itself become almost extinct.)

THE RECLUSE OF DOWN

Charles Darwin (1809–1882) was born to a moderately wealthy English family. As a child he was considered mentally inferior. At Cambridge University, where his father thought the boy was studying for the ministry, good-time Charlie spent most of his days with his buddies, drinking, hunting, and playing cards.

After the voyage of the *Beagle*, Darwin married his first cousin, Emma Wedgwood, granddaughter of the famed (and quite wealthy) pottery maker. He and Emma lived in near seclusion at his home, Down House, in the Kent countryside south of London. For forty years he kept to a clockwork schedule, interspersing walks around his private grounds with scientific work in his study and—his great pleasure—backgammon games by the thousands with Emma.

All his books were commercial successes, from the first, a chronicle of his adventures at sea called *The Voyage of the Beagle*, to the last, a page-turner entitled *The Formation of Vegetable Mould Through the Action of Worms*. (Honest.) Presaging the twentieth-century field of sociobiology, which studies how evolution has shaped the behavior of animals and people, Darwin even wrote a book on how human emotions are governed by evolved traits of the brain.

EVIDENCE OF EVOLUTION: WHAT DARWIN SAW WHILE REGAINING HIS LAND LEGS

There are three main pillars of evidence that Darwin recognized during his travels and that propelled him to seek an explanation.

1. The diversity of life. As we saw in Chapter 1, there is a mind-boggling variety of living things. In essentially every ecological niche available on Earth, there is some species that seems custom-built to live there. What can have produced so many millions of different life forms and "designed" them to fit their habitats so well?

2. The similarity of life. Even though there are lots of superficially different forms of life, there are many striking similarities among large groups of species. Darwin saw the similarities in gross anatomy, and you can check out the same thing. The next time your cat or dog is handy, feel its bones. For every bone in your body, you can find an exact counterpart in your pet. The shapes and sizes differ, but that's all. For some reason, the basic plan is the same. A giraffe has the same number of neck bones as a mouse. The giraffe's are just bigger. You could make just as good a giraffe with many more neck bones, but for some reason, that didn't happen.

Since Darwin's day, biology has extended the known areas of similarity to the smallest anatomical details. Nervous systems, for example, have the same kinds of cells and the same electrical and chemical methods of relaying signals in every species from worms to humans. Not just the same methods, in fact, but the very same chemicals. How come?

Go even deeper and check out the genetic codes used by all species on Earth. Turns out there is only one code. As we mentioned in Chapter 2, the code is arbitrary and many other codes would work just as well. If each species were created independently, why don't they have inde-

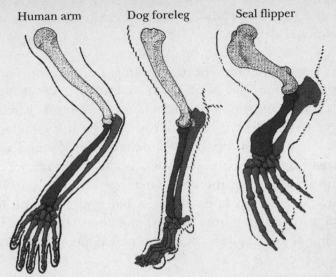

Human arm   Dog foreleg   Seal flipper

7.2 Present arms! The forelimbs of mammals all have the same basic skeletal structure—one upper arm bone connected to two lower arm bones at an elbow and then a bunch of little bones that form five fingers.

Cockroach

Esophagus   Crop Gizzard

Anus
Rectum
Intestine
Mandibles
Salivary glands

Nematode
Mouth   Anus
           Intestine

Earthworm   Intestine
Pharynx   Crop

Mouth   Anus
     Esophagus Gizzard

Snail
Radula
Mouth   Anus
          Intestine
    Liver Stomach

Rabbit   Salivary glands
              Pancreas
                      Caecum
                      Rectum

Teeth   Anus
Esophagus   Small
                    Intestine
Liver
        Stomach Large
                 Intestine

7.3 Gut reactions: Fundamental needs (digesting food and absorbing the nutrients) are met in much the same way by almost all animals, with a gut that runs from the mouth to the anus.

pendent genetic codes? Why is there only one genetic code on Earth?

3. *The fossil record*. The third thing Darwin noticed on his voyage was that when he looked at the layers of rock in a formation (he would have loved the Grand Canyon), there were different kinds of fossils in each layer. Geologists had already established that the rock strata were formed as sediments deposited on the bottoms of ancient seas, burying the bones that lived in long-vanished ages. And Darwin knew that the oldest were at the bottom and younger ones toward the top. That's the way sediment works—just like the newspapers stacked in the garage.

What struck Darwin was how the fossils differed through the layers. The lower and older, the simpler the organisms. The simplest extinct animals appeared only in the oldest strata and were absent from any higher position. More complex species made their first appearances in middle layers, persisted through a few succeeding intervals and then disappeared. The fossils most closely resembling currently living forms could be found only in the uppermost rock layers. For example, you never find horse fossils among those primordial crawlers of the ocean bottoms—trilobites, and, as long as Jimmy Hoffa is still missing, you'll never find human bones in the same layer as dinosaur bones.

The logic was inescapable. Something was causing species to change over time, causing some of them to become more complex—and the whole lot to become more diverse—as the eons passed.

NATURAL SELECTION: THE BLIND WATCHMAKER

Some people challenge the principle of evolution on the ground that nature is such a marvelously intricate machine that it could not have arisen by chance alone any more than a heap of gears and springs (or should we say quartz crystals and batteries) could suddenly assemble themselves into a watch. They are right. Evolution does not work by chance alone. It relies on a very powerful force—natural selection—that is quite nonrandom, quite deliberate. But chance *is* a partner in the process. Here's how Darwin reasoned it out:

Phenomenon No. 1

He noticed that most species have many more babies than are needed to maintain their population. To keep a species at a steady state, each pair of parents needs to produce only two surviving offspring. For example, if all the hundreds of progeny of just one pair of houseflies survived and reproduced and all *their* precious maggots grew up and had still more babies, in only one summer the world would become just too disgusting. In reality, nearly all flies die in infancy or youth because there isn't enough maggot food to go around. The world does not offer unlimited resources to any species.

Because more babies are born than can possibly survive, they (or their parents, in the case of humans and other species with long periods of dependency) are forced to compete for resources.

Phenomenon No. 2

The babies of any given species are not identical. There

are tiny differences among them. Some humans, for example, have big ears; others, small ears. The differences may
be anatomical and easily seen, or they may be biochemical
and hidden from view. A human born with slightly more
potent digestive enzymes, for example, might be able to
extract nutrients from Cheez-Doodles®.

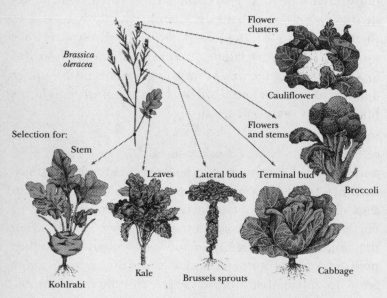

*7.4 Cutting the mustard. The genetic variability of
wild mustard plants led to the creation of new vegetables,
all belonging to the same species.*

Inevitable Result of the Two Phenomena

Those offspring that happen to possess differences that
give them an advantage in their environment are more likely
to survive than their less fortunate siblings. In an environment where Cheez-Doodles® were the main food source, for
example, those of us with those special enzymes should be
more likely to survive than other, less fortunate folk.

So what? So, because if we live to adulthood, we're more likely to have babies of our own. And they'll inherit the Cheez-Doodle®-using enzymes. Pretty soon the environment naturally will have selected and favored so many of us with that advantage that the species will have changed—evolved.

Darwin called this process "natural selection" to distinguish it from the kind of selection that breeders of plants and animals had long carried out. It was well-known that all the breeds of cats, dogs, farm animals, and what-have-you were the result of people selecting animals with traits they liked (longer fur, shorter legs, bigger udders, propensity to herd sheep, etc.) and breeding only them. After several generations of this unnatural selection, a person could cause dramatic changes in both anatomy and behavior.

So, in a way, all Darwin added to this knowledge was the notion that nature does the same kind of selecting (by means of Phenomena 1 and 2) but at a pace so slow that no one could notice it happening. As he further pointed out, nature has had a lot more time to keep selecting than, for example, the members of the pigeon-fanciers club that Charlie hung out with sometimes. Nature, Darwin said, could accumulate lots of tiny differences over millions of generations and produce not just a pigeon with fancier feathers but pigeons in the first place—taking one species and accumulating so many tiny differences in it that the result is so different that it really qualifies as another species. (Fiction writers can only fantasize about flying monkeys; nature may already have tested a prototype and nixed the project.)

MUTATION: EVOLUTION'S CRAPSHOOT

Okay, so there are tiny differences among offspring. What makes the differences? The answer is: changes in the genetic code that one generation passes to the next, changes in the DNA carried by sperm or egg.

Darwin didn't know about this. He never heard of DNA. He never even heard of genes. Nobody had in those days, even though Gregor Mendel was doing his pea-breeding experiments at the same time Darwin published his greatest book. The Austrian monk published long before Darwin died but, as far as anybody knows today, word of his discovery of inheritable factors never reached Down House. Darwin simply saw that there were differences, assumed there had to be a natural cause for them, and that the cause was passed from parent to offspring.

Now, as it happens, there are two ways the genetic code can be relayed from one generation to the next: the boring asexual way (favored by bacteria and algae) and the much more interesting sexual way (highly favored by just about everybody else).

One of the great advantages of sexual reproduction (felt long after the initial excitement wears off) is that it allows sorting and mixing of many subtle biological characteristics. (The advent of sexual reproduction actually caused evolution to speed up.) Here's what happens: Just before the gonads make sperm or eggs, they actually move genes from one chromosome to another in random patterns—cutting a DNA segment here and pasting it there. As a result, the half-set of chromosomes that goes into the sperm or the egg is not identical to either half-set that exists in the cells of the parent's body. For example, let's say Dad had his gene for dimples on the same chromosome with his gene for long fingers. In his sperm, however, those genes may become separated and any given sperm may acquire one gene but not the other. The combination of dimples and long fingers will be lost in his child.

Children inherit all their genes from their parents, and so clearly resemble them, but only in certain features. This is why people say you and your siblings all look something like your mother or father but you all don't look exactly alike. Not only does this make it easier to remember who is who (imagine how confusing it would be if we couldn't tell one person from another), it generates the diversity on which natural selection can work. If the combination of dimples and long fingers were an advantage in some environment, the loss of that combination could be bad for those who don't inherit it.

There may be another reason why you aren't exactly like your relatives: You could be a mutant. No, you have not just been insulted. Human beings, after all, are mutant apes. Mutations are spelling changes within a gene. Sometimes just one letter in the sequence is altered, resulting in a different amino acid occupying that position in the resulting protein. (We covered this stuff in Chapter 2. Remember?)

Mutations like this can be caused by radiation, by extraneous chemicals that gum up DNA replication, or by inadvertent errors in the DNA-copying process. These things can happen anywhere in the body (and may cause cancer if they do) but the only changes that could directly affect the next generation would be those that happened in the gonads, altering the genetic messages passed on to the eggs or sperm. The changes are completely random. There's no telling which gene or genes may be involved.

A tiny mutation can, of course, make a big difference in the way the resulting protein works in cells. It might, for example, shift the time in a lifespan at which bone growth stops, resulting in the individual being larger or smaller than usual.

There is another kind of mutation that can happen, too. Remember the genetic gibberish segments that interrupt most genes? (That was in Chapter 2 also.) Here's where they may provide their evolutionary benefit. Among the kinds of things that gonad cells may do when they are manufacturing half-sets of chromosomes for sperm or eggs is to shuffle segments of genes around. They might even take a segment from one gene and stick it in the middle of a completely different gene.

Now, this, too, is a random process. When it's doing this, the cell doesn't know where one gene starts and another ends. And it doesn't know where the gibberish segments are. But if the lengths of gibberish are very long, which they often are, the odds are high that the breaks in the DNA chain will be in the gibberish. In other words, the odds are high that the part of the gene being shuffled contains one complete good stretch plus a little gibberish on each end.

Why would this matter? Excellent question! Because, as it happens, the good segments encode complete functional units of proteins. They are modules. You could compare it to a stereo system that has a CD player module, an amplifier module, and speaker modules. Break the stereo into functional units and shuffle them with, say, the modules of a personal computer and you might end up with a useful new combination. Maybe the stereo would acquire a monitor that displayed the music in pretty colors and patterns. In an environment that favored this sort of thing (call it "psychedelic" and it might sell), the new combination could prevail. But if the shuffling were not modular, you might wind up with, say, the CD player's laser pointlessly aimed at the speaker. Note, however, that even modular shuffling is still random and, you could also wind up with an intact CD player sending its signals to an intact but completely unresponsive keyboard.

Evolution is like that. Most of the random combinations of modules are no good. But a few are. What, for example, if the gene for hemoglobin (which takes up oxygen from air in the lungs) got a new module that made it able to grab oxygen from water in the lungs. Maybe humans would suddenly be able to live underwater. (Would they find airfront property desirable?) Whatever.

One way or another, gonad cells are constantly generating little random differences in the genes of the next generation. Some are lethal, often killing the embryo long before it is born. Some have no effect one way or another. A few are beneficial and, under the pressure of natural selection, they enable the individual to prosper in its environment—or some other environment—as its ancestors never could.

For example, let's suppose that the world of the chicken suddenly became flooded and the hapless birds had to float on the water and paddle about. Because most chickens don't have much webbing between their toes, most chickens wouldn't move very fast. But if one odd chicken happened to have more webbing (a trait that might have been a disadvantage on land), it could paddle faster—maybe escape predators more easily. Guess which chicken would be more likely to survive and pass on its genes. Or, to make up another example, suppose it's a few million years ago and the first apelike creatures to walk upright are having a hard time balancing on just two legs. Their children just might inherit tiny, random variations in the shapes of their knee and hip joints. In the trees or walking on all fours, these variations wouldn't come to much, but under the new system of trying to walk upright, some of the kids might have inherited a more stable mechanical arrangement for support. Not only could they make more trouble for their parents, they would be better able to use their front paws as hands.

THE EVOLUTIONARY TWO-STEP
Just to recap, evolution by natural selection is the result of two different but complementary phenomena.

STEP ONE: Small changes in the genes (mutations, for example) are generated whenever gonads make sperm or eggs. This is a random event.

STEP TWO: Environmental pressures (including competition from others for limited natural resources) determine which (if any) of the small changes will prove advantageous. Individuals helped by those advantages are more likely to live and more likely to have more offspring. The offspring inherit the advantages. This is decidedly not a random event.

THE ORIGIN OF SPECIES: IT'S NEW AND IMPROVED!

How does natural selection create whole new species? The answer is "reproductive isolation." Some members of the original species must become reproductively isolated from the other members. If individuals bearing the new and improved trait can interbreed with all members of the species, the gene for any improvement will gradually spread throughout the whole species. This takes many generations but in time the species as a whole changes. If this were the way evolution always worked, there would be only one species on Earth today and it would embody all the changes that ever came along. There would be no biodiversity.

The more usual course of events is for one ancestral species to split in two, giving rise to a second, descendent species, even as the ancestral form continues living the old way. For this to happen requires some form of reproductive isolation, some way in which the new model is prevented from interbreeding (mixing its new gene) with the old model.

Here's a simple example: A storm blows a small flock of birds to an uninhabited island far out in the ocean—much farther than those birds can fly on their own. When the birds get there, they find an environment with no other birds of their species. Now the mutations that happen to the island birds will cause them to change but cannot affect the same species back on the mainland. The two populations are reproductively isolated. After a long time, the island birds become so different that even if they did meet their mainland ancestors, they wouldn't give each other the time of day.

Reproductive isolation can also happen on a very gradual scale as when continents drift apart, or rapidly as, for example, when a meandering river changes course and cuts through the middle of the habitat of a species that cannot cross rivers. Or the organism itself can do the separating. A band of migratory animals might, for example, strike off into a new territory and lose contact with its ancestors.

This is how the various human races developed, each in a different part of the world and having so little contact with the other that differences in skin color and facial form developed. But we know that this reproductive isolation was not complete because all human races can interbreed and produce perfectly healthy children. The isolation was lost long before humans split into separate species. In fact, there is now so little isolation among human races that racial differences are blurring in many parts of the world. Over a very long time in human terms, but a blink in nature's, this could mean the people of the future will be of one blended race.

In many animal groups, reproductive isolation can also happen as a result of what seem to be tiny changes in instinctive behavior. Among insects and birds, for example, males can't get to first base with females unless they know exactly how to wiggle their wings or whistle their tunes or what have you. Performance of and appreciation for courtship rituals are hard-wired into the brains of the opposite sex. That means they are genetically governed. If there is a mutation in the genes involved, the animal's mating dance may change.

So a male bird may be born with a weird courtship dance. To most of the females, he's a dork. But if there's a female who isn't too put off or who actually likes the performance, she will mate with him and produce off-spring who may all find the mutant courtship ritual sexy. These birds look at first exactly like their ancestors. But because they are now their own breeding population, behaviorally separated from the others, they are an instant new species, and any random mutations that crop up in their genes will be theirs alone. In time they will become very different.

In a very long time, they could evolve so much that they no longer qualify as birds but become some entirely new kind of animal.

THE ORIGIN OF LIFE: PRIMORDIAL SOUP DU JOUR

This probably ought to be the shortest section of this book because, frankly, nobody knows how life began. It may be the second most challenging mystery facing science. (The origin of the universe is the biggest baffler.) But we won't let that stop us.

Darwinian evolutionary principles (plus all the subsequent findings of biology) tell us that all living things must be descended from a simple, one-celled organism that lived around 3.8 to 3.9 billion years ago. Which makes life even older than George Burns. The Earth itself didn't come into existence until about 4.5 billion years ago, and didn't cool off enough for water to exist in the liquid state for about 600 million years. One clue to life's antiquity is that in Greenland and southern Africa there are some fairly obvious fossils that look like bacteria in rocks that formed 3.5 billion years ago.

That early date suggests that life arose on Earth just as

soon as it chemically and physically could. In other words, it was not the result of some wildly improbable coincidence that requires billions of years of dice-throwing to hit the jackpot. Instead, life may have been the inevitable product of chemical and environmental conditions. There are numerous studies showing that the molecules needed for life will form spontaneously under conditions on the primitive Earth.

The most famous experiments showing this were done in 1953 by the American biochemist Stanley Miller. He took three gases that were common to the ancient Earth's atmosphere—hydrogen, ammonia, and methane—sealed them in a flask with water at the bottom, and zapped electric sparks through the mixture of vapors. The idea was to simulate lightning in the primordial atmosphere and absorption of the resulting material in the water.

After a week of sparks, the water had turned brown. The electrical energy had broken up the gas molecules and regrouped them into new and more complex molecules. When he analyzed the water, Miller found a mixture of organic compounds like those previously thought to be created only by living cells. In the murky soup, he found several different amino acids (the relatively complex building blocks of proteins), fatty acids, some sugars, and urea. Other experiments showed many of the same substances were created if the gas mixture was simply heated or exposed to ultraviolet light of the sort that comes from the sun. In other words, many of the molecules called "organic compounds" because they were thought to be produced only by living organisms can be created under entirely inorganic, or nonliving, conditions.

Other experiments by the American biochemist Sidney Fox found that when the amino acids from a Miller-type experiment are heated, they link themselves into proteins. And he found that if he put the proteins in water, they spontaneously assembled themselves into microscopic spheres. The spheres are definitely not cells, but the experiment showed that proteins created under inorganic conditions have the intrinsic ability to link themselves into larger structures. This is another example of the principle of self-assembly (Chapter 3) that underlies so much of what makes life possible.

More recently, similar experiments done by Cyril Ponnamperuma, a Sri Lankan-American biochemist, have shown that the subunits of DNA also form spontaneously under nonliving conditions.

Tally up these results and you will see that under the right conditions (all of which existed naturally on the ancient Earth), ordinary chemical reactions can create all the major building blocks of a cell. These are fats (the main component of membranes), carbohydrates (energy storage depots), proteins (enzymes and cellular structures), and nucleic acids (to store the genetic code).

Thus, it is very tempting for biologists to imagine that the ancient Earth must have harbored countless places—Darwin called them "warm little ponds"—where these molecules were produced and, somehow, came together to form the first cell. Actually, biologists suspect this must have happened many times in many places and that there were innumerable "first" cells.

But there is a very large stumbling block. Even the simplest known bacterial cells rely on a mechanism that seems unlikely to be able to come into existence by itself. Bacteria, like all modern cells, have protein enzymes to copy their DNA so they can reproduce. They get the enzymes by reading the genetic messages encoded in their DNA. In other words, you need specific enzymes to make new DNA and you need DNA to make the enzymes. You can't have one without the other. The notion that both DNA and the enzymes to replicate DNA could have spontaneously formed in the same place is wildly improbable.

In recent years, however, a new discovery has suggested a way out of the conundrum. Scientists have discovered that certain forms of RNA can act as enzymes that carry out their own replication. RNA, or ribonucleic acid, if you remember from Chapter 2, is the molecule that carries a copy of DNA's genetic message out to the rest of the cell. Thus, RNA is a molecule that can both carry genetic messages and act as a catalyst, or enzyme, causing a copy of itself to be made. Could the first cells have used RNA and not DNA? Many biologists find the prospect tantalizing and imagine that life on Earth was once an "RNA world."

One of the enzymes that might have evolved later on might have been "reverse transcriptase." It exists today and is used by certain viruses to copy their RNA genes into DNA and then to splice those copies into the DNA of an infected cell. That's how the AIDS-causing virus works. But, as it happens, RNA breaks down easily. So if the reverse transcriptase enzyme happened to arise, it would have immediately upgraded the reliability of the inner workings of living things by converting the RNA genes (of whatever organism it arose in) into the much more durable form of DNA. As hardware goes, DNA is so much better for the purpose of storing genetic information (software) that once it arose, it is difficult to imagine any RNA-based organisms surviving. That could explain why there aren't any around.

Whatever the first cell was, it must have been a simple

ET IS HOME

Extraterrestrials, by all odds, are real. Biologists generally agree on that because there are almost certainly trillions of other planets in the universe. Even if conditions are right for the origin of life on only one in a million, that would mean there surely must be life on millions of other planets. And on some of them, there is likely to be intelligent life. (Some even argue that it exists on Earth.)

7.5 E.T. far from home.

They base this belief on the fact that there are millions of galaxies and millions of stars in each galaxy. And that many of those stars surely have planets—some of which are at the right distance from the star to get the solar energy life needs and hold liquid water. Work out all the probabilities and it turns out that there are probably millions of planets out there with intelligent life.

Unfortunately, however, we will probably never meet them because to reach even the nearest such planets at the speed of light would take a lifetime or more.

thing that couldn't do much. But it wouldn't have had to. As the only game in town, it would have had no competition and no enemies. As long as it could reproduce, it could transmit the spark of life with various added attributes as evolution got going.

THE BEGATS OF BIOLOGY: FROM THE AGE OF ALGAE TO THE AGE OF BEAVIS AND BUTT-HEAD

Life began in the water. It could not have survived on land because in those days there was no oxygen in the atmosphere and, therefore, no ozone layer (it's a form of oxygen) to filter out the sun's lethal ultraviolet rays. Water, however, shields against UV, so that's where all the earliest critters were stuck. Three billion years ago all of Earth's land surface was bare rock and rubble. But possibly every object in the ocean that sunlight could reach was covered by a scummy mat of microbes, possibly something like today's bacteria. They made a living by carrying out photosynthesis, capturing and using the sun's energy.

Bacteria ruled the Earth for at least half the time life has existed on this planet. They began at least as far back as 3.5 billion years ago (maybe closer to 3.9 billion years ago) and their reign was unchallenged until about 1.8 billion years ago. That's when the first of a new kind of cell emerged. The newcomers were still single-celled organisms (algae, mainly) but their anatomy was fundamentally different. Their DNA was protected inside membranes (as a nucleus), and they contained other membrane-enclosed organelles such as mitochondria.

These advances made possible all the biodiversity that would follow.

In other words, life itself originated very quickly on the primordial Earth, but it took a very long time to evolve the next big step—the advanced-model cell. Biologists today call the simpler cell type prokaryote ("pro-carry-oat," from the Greek for "before kernel, or nucleus") and the upgraded type eukaryote ("you-carry-oat," or "true nucleus"). Today the only prokaryotes are the bacteria and closely related organisms. All the others—protozoans, fungi, plants, and animals—are eukaryotes.

Even after the first eukaryotic cells evolved, however, all living things remained microscopic in size. Yet their contribution to the subsequent evolution of life was enormous: they produced the oxygen that gradually made it possible for still more advanced organisms to arise. They released this oxygen to the water, and some of it escaped to the air.

Sometime between 800 million and 600 million years ago, some of the single-celled organisms stopped going their separate ways after cell division and, instead, stayed stuck together, forming the first multicelled animals. Instead of each cell having to carry out all life functions for itself, they could begin to specialize. Some changed shape to be more efficient at feeding; others, to maximize the effectiveness of offensive structures, such as stingers, or defensive structures, such as shells. Multicellularity allowed organisms to become far better equipped for survival in specific habitats and ways of life.

As a result, starting around 540 million years ago, the Earth witnessed the most exuberant burst of evolutionary diversification known. The so-called Cambrian explosion

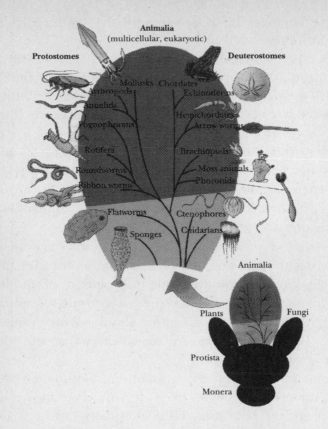

7.6 The family tree. Schematic showing the probable evolutionary relationships among major groups of living things. Humans would be nearest the frog in this chart.

(named for Cambria, also know as Wales, the first place the fossils from that period were recognized) filled the oceans with life—creating many of the body plans known today, and, possibly, a global biodiversity greater than that of today. There were sponges stationary on the bottom, trilobites with eyes and legs, crawling about, hard-shelled mollusks jetting through the water. And much more. Nobody knows for sure why the Cambrian explosion happened, but one theory is that the amount of oxygen in the air and water, which had been rising slowly for millions of years (thanks to the photosynthesizing bacteria and algae), reached a high enough level. When the oxygen level is low, cells on the surface or even in gills or lungs cannot absorb enough to supply large internal volumes of tissue. Animals must stay small. But once the oxygen concentration gets high enough, it becomes possible to sustain large bodies with the amount that can be taken in by gills or lungs.

Shortly after the Cambrian explosion, the first plants invaded the land and fish evolved in the sea. Some fish evolved, maybe around 350 million years ago, into amphibians—creatures with lungs to breathe air but still needing to go back to the water to breed. Some of these amphibians evolved into the first reptiles about 300 million years ago.

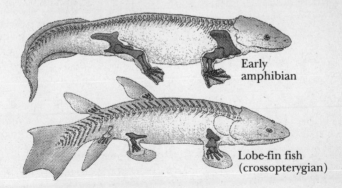

Early
amphibian

Lobe-fin fish
(crossopterygian)

7.7 A fish out of water. Skeletal analysis of the lobe-fin fish shows it needs only slight modification of its bones to resemble an early land-dwelling amphibian.

As everyone knows, some of the reptiles evolved into dinosaurs around 225 million years ago. But at about the same time, as very few people know, some other reptiles evolved into the first mammals. A bit later still other reptiles evolved into birds. Reptiles have been very, very good to our modern biodiversity.

GEOLOGICAL HISTORY OF LIFE

EON	ERA	PERIOD	BEGAN	MAJOR EVENTS
Archaen			3.9 byr*	**Origin of life**; age of bacteria
Proterozoic			2.5 byr	First eukaryotes (complex cells with nuclei); **early animals appear**
Phanerozoic	Paleozoic	Cambrian	600 myr*	**Explosion of evolutionary diversification**; most animal phyla arise; algae dominate waters
		Ordovician	500 myr	Much diversification within animal phyla; **first jawless fishes** mass extinction closes period
		Silurian	440 myr	First bony fishes; **plants and animals invade land**
		Devonian	400 myr	Fishes diversify; first insects; **some fish evolve into amphibians**; mass extinction closes period

EON	ERA	PERIOD	BEGAN	MAJOR EVENTS
		Carboniferous	345 myr	Vast forests of ferns and cycads cover land; **some amphibians evolve into reptiles**
		Permian	290 myr	Continents aggregate into Pangaea; reptiles diversify; insects diversify; major mass extinction closes period
	Mesozoic	Triassic	245 myr	Continent begins drifting apart; **some reptiles evolve into dinosaurs, and other reptiles evolve into mammals;** mass extinction ends period
		Jurassic	195 myr	Dinosaurs diversify; **some dinosaurs evolve into birds;** Steven Spielberg gets idea for movie
		Cretaceous	138 myr	Most continents widely separated; dinosaurs continue diversifying; first flowering plants evolve; mammals diversify a little; **big Blam causes yet another mass extinction**
	Cenozoic	Tertiary	65 myr	Continents near current positions; major diversifications of birds and mammals;

EON	ERA	PERIOD	BEGAN	MAJOR EVENTS
				and, yes, another mass extinction
		Quaternary	2 myr	Climate turns cold with repeated glaciations; **some apes evolve into people**; many large mammals become extinct; people begin to cause what could be another mass extinction

*byr = billion years ago, myr = million years ago

IT LOOKS LIKE A DUCK AND WALKS LIKE A DUCK

When some of the more primitive type of cell (prokaryote) evolved into the more modern type (eukaryote), one of the key steps was the acquisition of mitochondria, the organelles that package the cell's energy into the form of ATP (Chapter 2).

Mitochondria are the size and shape of bacteria. Mitochondria multiply within cells by dividing. They carry DNA of their own—and, like bacteria, have no nucleus to keep it in. And like those microbes we tend to dispise, mitochondria have their own ribosomes to read their genes and make their own proteins.

Many biologists suspect that the mitochondria in all our cells are actually bacteria that took up symbiotic residence billions of years ago. Without them we'd be dead. At the level of our cells, we are almost surely part germ.

SHOOTOUT AT FOSSIL GAP

Some people don't accept evolution as a fact because they say there are glaring gaps in the fossil record. If reptiles really evolved into mammals, or into birds, they say, why aren't there any fossils of transitional species? Where are the bones of a creature that's half reptile and half mammal—a repmal or mamtile? Likewise, they will ask, can anybody point to a fossil that's intermediate between a reptile and a bird?

These claims frustrate biologists because there are lots of transitional fossil forms. There are gaps in the fossil record, to be sure. Fossils are rare things that form only where the geology is just right—which may not be where the transition happened. But still, they are there. For example, among the reptile fossils from before the dinosaurs are some of what are called mammal-like reptiles. In many ways, they are intermediate. And there is *Archaeopteryx*, a small creature that had feathers on its body (including wings, which had claws) and had teeth. You can think of it as a feathered dinosaur that was halfway to being a bird, or as a bird that retained some toothy relics of its dinosaurian ancestry.

Perhaps the evolutionary transition best documented by fossils is that from ape to human. There are dozens of fossil specimens representing at least four evolutionary steps plus some gradual shadings from the very ape-like "Lucy" (*Australopithecus afarensis*, who lived nearly four million years ago) through *Homo habilis* and *Homo erectus* to archaic forms of *Homo sapiens* to modern humans. The transitions are often so gradual that experts debate where to draw lines.

THE BIG BLAM! AND OTHER EXTINCTIONS

About sixty-five million years ago something awful happened. A big rock fell out of the sky—a *really* BIG rock, at least six miles across. It was falling at a speed that may have been around 50,000 mph—a typical speed for a meteorite. It slammed into the Earth with such force that it vaporized itself and blasted out enough dust to shroud the planet in a dark cloud for months or maybe even a year or more. Not only did this deprive plants of sunlight, it deprived animals of the solar energy that plants capture. Without plants, animals are dead meat.

That's when the dinosaurs became extinct. There is evidence the dinosaurs were dying out before that, probably due to climate change, but the Big Blam finished off the last of them. And not just dinosaurs. Countless other species also perished. The fossil record is most complete for marine animals and it shows that two-thirds of their species became extinct during this event.

In western North America, the number of flowering plants (the dominant form on Earth today) crashed sixty-five million years ago. Immediately after, the plant fossil record shows, there was a sudden surge in the fern population. The fern "spike" is consistent with a darker, cooler climate that may have prevailed for a few years after the impact. Then many of the flowering plants bounced back, perhaps sprouting from seeds that had survived in dormancy.

The mass extinction—or, more precisely, the sudden change in the fossil assemblage—marks the end of the Cretaceous period and the beginning of the Tertiary. This most famous mass wipeout (though not the largest) is known as the K-T extinction—K for Cretaceous (that makes sense, doesn't it?) and T for Tertiary.

With the dinosaurs and many other species out of the ecological picture, the door was open for a new burst of evolution—just as in a small town where the only bakery closes, there is opportunity for a new one to move in. Except that evolution doesn't know how to retrace its steps. Instead of creating new dinosaurs, it created new mammals. The mammals were there in the days of the dinosaurs, but they were shut out of all the good ecological niches. With the amazing hulks gone, there was no competition for new experimental versions of mammals. In a mere twenty million years more, Earth regained about as much biological diversity as it had before the rock fell.

Mass extinctions have devastated life on Earth at least five times, the K-T event being only the most recent. The worst one closed the Permian period 245 million years ago, erasing 96 percent of the marine species living then. Life, in other words, nearly ended. Again, it took many millions of years to create new biodiversity. But Steven Spielberg should be grateful. It was that event that simultaneously opened the way for the first dinosaurs—and the first mammals, some of which evolved into fans of *Jurassic Park*.

NOT WITH A BLAM BUT A WHIMPER

Extinction can happen without big rocks falling out of the sky—and usually does. In fact, over the past 600 million years, each of the ten major periods of geologic history has seen more than 99 percent of its species vanish. So why is biodiversity still so great? Because for each species that bites the dust, another eventually rises up. Out of a thousand reptile species that croaked in the

LANDMARKS
These Magic Moments

1859—Charles Darwin (1809–1882)
English naturalist. Publishes *On the Origin of Species*, asserting that evolution explains biodiversity and that it happens gradually by means of natural selection. But can't explain the cause of heredity. Never hears of Mendel.

1865—Gregor Mendel (1822–1884)
Austrian monk. Discovers that heredity is transmitted in discrete "units," later called genes. Asserts each trait is governed by two genes, one inherited from each parent. Never hears of Darwin.

1900—Several scientists independently rediscover Mendel's work but hold that genes cause big changes, not the gradual ones Darwin invokes.

1920s to 1940s—Theodosius Dobzhansky (1900–1975) Russian-American geneticist. His fruit-fly breeding experiments show that most genes govern very small events. Thus gene changes could, after all, be the agents of Darwinian evolution.

1944—George Gaylord Simpson (1902–1984)
American paleontologist. Publishes *Tempo and Mode in Evolution*, linking Darwin's theories and the genes of Mendel and Dobzhansky to the fossil record. Shows how all the pieces of the puzzle fit together.

LANDMARKS
(*continued*)

1979—Luis Alvarez (1911–1988)
American geophysicist. Reports evidence of a major asteroid impact that could have disrupted Earth's climate and caused the mass extinction that wiped out the dinosaurs. Triggers new appreciation of the role of extinction in evolution.

Permian extinction, let's say, one survived to be the ancestor of a thousand dinosaur species. And, in turn, although all of them were eventually wiped out, they did not perish before giving rise to birds and all their colorful diversity, which exceeds a thousand species by far.

Evolution continually generates new experiments in anatomy and behavior, then sends them out to see if there is any way they can make a living in the richness of Earth's many habitats. But this is not a quick process. It does not occur rapidly enough to keep up with the mass extinction being waged by one of evolution's newer creations, *Homo sapiens sapiens* ("doubly wise," talk about ego!) Because of human decisions and indecision, species are being wiped out at a pace far above nature-as-usual. Extinction may not be happening as fast as at the K-T boundary, but even if it takes several decades to match that level of destruction, Earth may become just as impoverished.

Evolution will replace the loss, to be sure, but we—or our evolutionary descendents—will have a very, very long time to wait.

SUMMARY

⏱ Charles Darwin based his theories of evolution on three main bodies of evidence: 1) the diversity of life 2) the similarity of life 3) the fossil record.

⏱ Natural selection results from two phenomena. First, virtualy all species produce far more offspring than can possibly survive to reproduce. Second, offspring are not identical. Small differences distinguish one from another. Thus, in the competition for resources, some will have the attributes that give them the edge in survival and others will not.

⏱ The basis of inheritance is genes. Random variations in gene structure that arise are called mutations. Most mutations are either harmful or inconsequential. But a few prove beneficial and their inheritors will be the ones to lead some members of the old species into a new evolutionary path.

⏱ Evolution explains how an early life form could have produced so much diversity, but it says nothing about how the first life form began. On this one, the jury is still out. There is good evidence, including fossils of bacteria, that the first life arose on Earth at least

3.8 billion years ago, relatively soon after the primordial Earth cooled below the melting point of rock.

Around 540 million years ago life underwent the so-called Cambrian explosion, the most exuberant burst of evolutionary diversification known. Not long after, the first plants invaded the land and fish appeared in the sea. Then came amphibians, reptiles, dinosaurs, birds and mammals.

Around 65 million years ago, a huge object slammed into the planet and wiped out the dinosaurs and a high proportion of all other species. That cleared the evolutionary stage for mammals to diversify, taking over many ecological niches once occupied by the dinosaurs. After many more episodes of global environmental change, apelike creatures arose, one of whom must have gotten reproductively isolated from the others and evolved into a species capable of studying its own extraordinary origin.

ABOUT THE
AUTHOR

BOYCE RENSBERGER is one of America's leading science writers and science editors, having put in more than thirty years at the craft, and has won numerous national awards for coverage of research, mainly in the life sciences. Twenty-three of those years have been at newspapers (starting at the *Detroit Free Press,* moving on to *The New York Times,* and now at *The Washington Post*). Rensberger was also senior editor of a monthly magazine that started out as *Science 81* and gradually became *Science 84.* He was head writer of the first season of "3–2–1 Contact!" a public television show on science for children. He has written three books besides this one, all explanations of scientific matters for the nonspecialist, none as much fun as this one. Although Rensberger holds an undergraduate degree in zoology from the University of Miami, that was a long time ago, and he says he has since learned far more in thirty years of interviewing the top experts in science for news stories.